A FIELD GUIDE TO
ANIMAL TRACKS

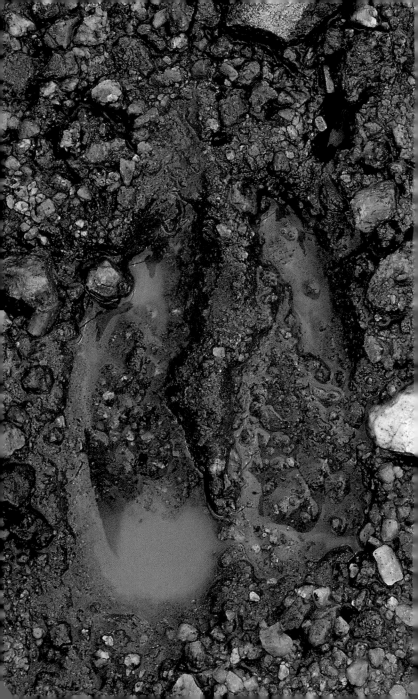

THE PETERSON FIELD GUIDE SERIES®

A FIELD GUIDE TO

ANIMAL TRACKS

THIRD EDITION

OLAUS J. MURIE AND
MARK ELBROCH

Illustrated by
OLAUS J. MURIE AND
MARK ELBROCH

SPONSORED BY THE NATIONAL AUDUBON SOCIETY,
THE NATIONAL WILDLIFE FEDERATION, AND
THE ROGER TORY PETERSON INSTITUTE

 HOUGHTON MIFFLIN COMPANY
BOSTON   NEW YORK

For information about permission to reproduce selections from this
book, write to Permissions, Houghton Mifflin Company,
215 Park Avenue, New York, New York 10003.

Visit our Web site: www. houghtonmifflinbooks.com

PETERSON FIELD GUIDES and PETERSON FIELD GUIDE SERIES are
registered trademarks of Houghton Mifflin Company.

Library of Congress Cataloging-in-Publication Data
Elbroch, Mark.
 The Peterson field guide to animal tracks / Olaus J. Murie and
Mark Elbroch — 3rd ed.
      p. cm.
   Rev. ed. of: Field guide to animal tracks / text and illus. by
Olaus J. Murie. 2nd ed. ©1974.
   Includes bibliographical references and index.
ISBN-10: 0-618-51743-x (pbk.) ISBN-13: 978-0-618-51743-5 (pbk.)
   1. Animal tracks — North America — Identification.
2. Animals — North America.   I. Murie, Olaus Johan, 1889-
1963.   II. Murie, Olaus Johan, 1889–1963. Field guide to ani-
mal tracks.   III. Title.
   QL768.E43 2005
   599'.097 — dc22   2005013108

All photographs by Mark Elbroch except as noted.

Book design by Anne Chalmers
Typeface: Linotype-Hell Fairfield; Futura Condensed (Adobe)

Printed in Singapore

TWP 10 9 8 7 6 5

Contents

PREFACE TO THE THIRD EDITION

OLAUS'S STUDY was a big room taking up nearly half the upstairs. On your right as you entered, a large desk filled the corner. A pair of north-facing windows gave good light on the desk top, which overflowed with papers, journals, and photos. A scrollwork inkstand stood at the back against the wall, carved by Olaus's father, a woodworker and immigrant from Norway. Between the door and the desk stood a metal table supporting the typewriter. My mother, Mardy, did most of the typing, but Olaus sometimes had to hunt-and-peck his way through a letter.

The inner walls of the study were of Celotex, the affordable wallboard of those days. The west windows lighted a long table. That too was loaded with books, papers, boxes, oils and watercolors and palette, but it could be cleared for special projects, such as analysis of sea otter scats and raven pellets or layouts of mammal or bird specimens. There were metal file cabinets and on the north wall a wooden cabinet for the herbarium. Bookcases held journals and a grand array of miscellany, and in the southeast quarter of the room stood the tall easel where quite often a painting was in progress.

We three kids had free access to the study, though we knew we could not mess with anything in there. One evening I was in there on some errand I've completely forgotten; Olaus was at the desk. I noticed a large sheet of scratchboard on the near end of the desk. A mink in bold India ink loped across its lower right corner, and above were track patterns. I asked about it.

"Might be a book someday," Olaus said. In a few words he told me what he had in mind. There would be tracks and scats and other signs.

I was entranced. That is no exaggeration. The leaping mink illustrating its own signs! And a page like that for every mammal.

"Birds too?"

Yes, as many as possible, and insects, amphibians, reptiles. Whatever and whoever leaves signs of occupation or passage. Olaus warmed to the subject, and now I learned why the stick tattooed by a sapsucker had been lying all this time on one of the bookcases, and the rodent-gnawed base of an elk antler, and the twigs incisored off by rabbit, deer, elk, beaver... These artifacts and many others had been taken for granted in my youthful mind as merely a part of the "data" that Olaus and my uncle Adolph gathered and talked about. They used that word "data" a lot. I learned later that it signified scientific accuracy, objectivity. Data were the raw materials upon which theory had to be built, and there was no shirking that.

What thrilled me about the prospective track book was the thought that tracks and signs could emerge from raw observation to appear in black-black ink on paper. And the animals in action, making their marks! I was, even at that early age, in love with signs that magically appear on paper.

And so the artifacts accumulated, the plaster casts and the blocks of tracks cut from dried mud, and photographs and boxes and jars holding scats. Big shelves in the basement became burdened with these things. When Mardy and Olaus moved, all that data went with them, to be housed in one of the unused cabins of the old dude ranch.

And then Houghton Mifflin: the acceptance, the contract, and the commitment; the long winter, Olaus making drawings, on a card table in front of the fireplace, and Mardy reading aloud. When I look through the stack of originals for the track book, I wonder how he, and Mardy reading, ever managed to bring it off, all those signs, all those animals.

One day Olaus and I were traveling across the Southwest. I was on army furlough; he was on a mission for the Wilderness Society. We came to one of those little zoos that service stations used to have as adjuncts to gas and tire-repair enticements. Olaus got out his soft, black-covered three-ring notebook, and we wandered past the few cages. Olaus stopped at one, and you know, I can't remember what animal was miserably penned there. I do remember glancing at the first few lines Olaus drew, noticing how quickly they limned the animal. Was it a bear? A coyote? Oh well, it's the lines on paper that I remember. I think that, speaking generally, his best artwork is in the line drawings.

On another occasion, in the Jackson Mountains of Nevada, looking for traces of desert mountain sheep, we stopped at bird tracks in the dust, and Olaus used some of our precious water to mix plaster and pour. I understood, and approved, but man, that was a thirsty day. And it was a day of careful study of animal sign.

On the dry, rocky ridges no track was perfect. The task was to distinguish between mule deer tracks and those of bighorns. It was finally decided that no sure sign of bighorns could be reported to the Biological Survey. The question remained unsettled.

There are times when marks of passage have to be given an educated guess, not a proclamation. Nature is not always crystal clear; Olaus was adamant on that point: honesty is the best policy. On the other hand, we curious mammals have that deep urge to decipher clues, to solve mysteries. It's one of the perennial fascinations in tracking—pushing the limits, using all our senses, trying for definitive answers.

In the fifty years since this guide made its first appearance, field biologists have amassed a treasure trove of new insights into the subtleties of animals and their ways. Mark Elbroch, a biologist who has been in the creative midst of these developments, brings that new knowledge and expertise to his thoughtful revision of *Animal Tracks,* so that Olaus's work may continue to be useful for many years to come. This revised and updated third edition will sharpen our senses, lead us into woods and marshes, deserts and Arctic tundra, and suburbs too, with fresh motivation, making us better naturalists.

—Martin Murie

OLAUS MURIE'S *Animal Tracks* was originally published in 1954, before I was born. Yet his work has remained a cornerstone in the blossoming field of wildlife tracking, and in wildlife research in which certain tracking skills are utilized in data collection. *Animal Tracks* is without doubt the most influential tracking guide to date. Murie reintroduced untold thousands, if not millions, of people to our amazing world, a world filled with uncountable signs of hidden wildlife. I hope this revision ensures that thousands more are exposed to Olaus's writings, accurate illustrations, and rare observation skills.

Animal Tracks is also a classic in nature writing and natural history. In his introduction Martin Murie paints us a vivid picture of the working Olaus, which together with the anecdotes sprinkled throughout the text helps the reader realize that Olaus Murie and this work are of an era past—a time when biologists were also naturalists, and naturalists were incredible observers. Yet Olaus was exceptional then, too. He actively *participated* in nature, captured badgers by their hind legs, dug up thirteen-lined ground squirrels, hunted, and trekked in true wilderness. He tempered excellent science with personal experience, while contributing much to research and wildlife politics as well as working tirelessly to protect wild creatures and wild places. His rhythmic writing and turns of phrase remind me of my grandfather, another great naturalist and observer.

Scientific knowledge is never static but rather an ever-evolving body of knowledge growing out of the works and discoveries of earlier scientists. Olaus Murie contributed vast amounts to our understanding of animal tracks and signs, which has allowed those who followed to build on that foundation and to further research additional species and topics. The most significant change in this third edition is in the branch of science called phylogeny, which is represented by the order in which the animals appear in the book and how they are grouped together. In theory, the first mammal to have finished evolving into the mammal it is today is first in the book, the second is second, until the very last. But the order reflects more than individual evolution, for evolution is not really a linear event. Groups of animals (cats, for example, or moles) split off in their evolution and appear before other groups of animals; evolution is happening on numerous fronts simultaneously. Visualize the many branches of a tree rather than a single timeline.

Genetic research has radically altered our work in phylogeny as well as the related taxonomy, which deals with the naming of species. The Virginia opossum is still the oldest mammal in North America in evolutionary terms and thus still appears first in the book. But thereafter begins the re-sorting and the many changes in both common and scientific names.

There is also a new Key to Tracks which I hope is more useful than the old when quickly referencing finds in the field. There is new natural history, and 91 new illustrations, not including the Key to Tracks. And of course the 105 color photographs are new. Yet each addition and change was made to blend with Olaus's existing work. This is Olaus Murie's book, and may it be useful to readers for another fifty years to come.

— Mark Elbroch

ACKNOWLEDGMENTS

FOR THE FIRST AND SECOND EDITIONS

This Field Guide is based almost entirely on my own field observations and the collection of plaster casts and other material that has been assembled since 1921.

However, I have also studied the extensive mammalian literature, which has been of great value, and a bibliography of some of this literature is given at the end of the book. Moreover, I have had the assistance of a number of persons, and without their help there would have been gaps in my information.

Dr. Hartley H. T. Jackson, with whom I was once associated in the United States Fish and Wildlife Service, many years ago suggested that I write such a handbook, and I began in a small way at that time.

Mr. Ernest P. Walker, assistant director of the National Zoological Park in Washington, D.C., together with members of his staff and the staff of the Woodland Park Zoo in Seattle and of the Fleishhacker Zoo of San Francisco, have all been of much help. With the help of attendants at the last-named zoo my son Martin Murie obtained tracks of the jaguar, which I could not have obtained otherwise.

The entomology department of the National Museum, the United States Fish and Wildlife Service, and the staff at Sequoia National Park have all given me assistance.

Many individuals have furnished valuable material: the late Francis H. Allen of Cambridge, Massachusetts; Dr. E. L. Cheatum of the New York Conservation Department; Dr. C. H. D. Clarke, Department of Lands and Forests, Ontario; Antoon de Vos, Ontario College of Agriculture; Lieutenant and Mrs. H. H. J. Cochrane, Canal Zone; Dr. James Zetek, in charge of the research station at Barro Colorado Island, Canal Zone; Warren Garst, Douglas, Wyoming; Luther C. Goldman, Texas; John K. Howard, who, with cooperation of the American Museum of Natural History of New York, made available to me film on Arctic hares; Dr. William J. Hamilton of Cornell University; William Handley of the National Museum; William Nancarrow, Alaska; Ivan R. Tompkins, Georgia; Dr. Robert Rausch of the Public Health Service, Alaska; Erwin Verity of Walt Disney Studios; and Dr. Frank N. Young of Indiana University.

My son Donald live-trapped certain mammals for study of their tracks. Howard Zahniser, my associate on the staff of the Wilderness Society, gave me pertinent suggestions. My brother, Adolph Murie, during his long sojourn in Alaska, assembled a mass of valuable data on such animals as the wolf and the wolverine, as well as ptarmigan and other species. This was all put at my disposal.

When I fell ill in May 1954 some illustrations were still needed to complete the manuscript. The warm response of the many people who were asked to help was overwhelming. To every one of them I give earnest thanks, especially to Mrs. William Grimes of the Massachusetts Audubon Society and Miss Frances Burnett of the Harvard Museum of Comparative Zoology for their helpful research; to Dr. William Burt of the University of Michigan Museum of Zoology and Mr. Bert Harwell of California for their many useful suggestions; to Mr. Richard Westwood of the American Nature Association for taking a personal interest in the project; to the New Hampshire Fish and Game Commission, which obtained from the Blue Mountain Forest Association the feet of a 206-pound wild boar; and to Mr. Ellsworth Jaeger of the Buffalo Museum of Science and Mr. George Mason of the American Museum of Natural History, who both offered their services in finding specimens and making the necessary illustrations.

The seven plates of track drawings were done on very short notice by Mr. Carroll B. Colby of Briarcliff Manor, and to him I owe my deepest gratitude, as I do to my artist friend and neighbor Grant Hagen, who furnished the final sketches for figure 152.

Finally, I appreciate the cordial and intimate relationship with Mr. Paul Brooks, editor-in-chief of Houghton Mifflin Company, and members of his staff who have taken a genuine interest in the manuscript and whose encouragement has been very stimulating. Among these, I give special gratitude to Katharine Bernard, Helen Phillips, and Anne Cabot Wyman.

Whether in the field of natural history or elsewhere, one does not accomplish a piece of work alone. It must at least in part draw on the experience of others. It is a pleasure to acknowledge such help, and to express my gratitude for such participation in my efforts.

O. J. M. 1954

FOR THE THIRD EDITION

Tremendous thanks to Martin Murie for his support of this project and for his literary contributions. Thanks also to Lisa White at Houghton Mifflin for providing this opportunity, and to Anne Hawkins for her support, enthusiasm, and keen eye for proposals and contracts. And a heartfelt thanks to Mark Allison and Stackpole Books for sanctioning the project and supporting this work in every way, including granting permission to use a few slides I'd published in *Mammal Tracks and Sign.*

The research for this revision was short; most of the work had already been completed for *Mammal Tracks and Sign* and *Bird Tracks and Sign.* Thank you to all who contributed to those books — their generosity supported this project as well. In addition, thanks so very much to Fiona Reid for sharing her resources and work on the revised *Field Guide to Mammals,* in order that the science in her book and this one match. Thank you also to Marcelo Aranda, who generously and quickly sent me slides of Central American animals, some of which appeared in his own excellent guide to tracks and signs of mammals in southern Mexico and Central America.

Thank you to Tiffany Morgan for her support in the early stages of the project and for accompanying and tolerating a "research" trip to Texas. Thanks to Nate and the Kempton family for hosting and sharing in several days of exploration in the Colorado Rockies. Michael Kresky shared a hot spot for harvest mice in southern California, and George Leoniak provided the space for me to lay out and complete the project in the peace of southern Vermont's hardwood forests — thank you.

There are many I count on for their continued support. Thank you to my parents, Victoria and Lawrence, my grandmother Elizabeth Gorst, and my great uncle and aunt Robert and Mary Cross; and to Keith Badger, Nancy Birtwell, Fred Vanderbeck, Eleanor Marks, Kurt and Susie Rinehart, Mike and Diane Pewtherer, Jonathan Talbot, Walker Korby, George Leoniak, Nate Harvey, Keely Eastly, Frank Grindrod, Paul Rezendes and Paulette Roy, Louis Liebenberg, Dr. Jim Halfpenny, Charles Worsham, Sue Morse, and Jon Young.

Thanks also to all the animals. A special thanks to the gray fox pups who supported me each day in the early stages of this project, and to the fishers and black bears who support me now at its closing.

Mark Elbroch

The legacy of America's great naturalist, Roger Tory Peterson, is preserved through the programs and work of the Roger Tory Peterson Institute of Natural History. The RTPI mission is to create passion for and knowledge of the natural world in the hearts and minds of children by inspiring and guiding the study of nature in our schools and communities. You can become a part of this worthy effort by joining RTPI. Just call RTPI's membership department at 1-800-758-6841, fax 716-665-3794, or e-mail (webmaster@rtpi.org) for a free one-year membership with the purchase of this Field Guide.

KEY TO TRACKS

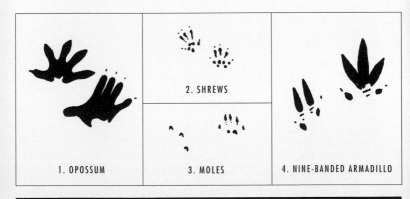

1. OPOSSUM   2. SHREWS   3. MOLES   4. NINE-BANDED ARMADILLO

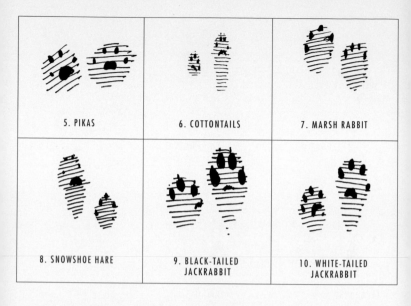

5. PIKAS

6. COTTONTAILS

7. MARSH RABBIT

8. SNOWSHOE HARE

9. BLACK-TAILED JACKRABBIT

10. WHITE-TAILED JACKRABBIT

11. APLODONTIA

12. CHIPMUNKS

13. MARMOTS, WOODCHUCK

14. GROUND SQUIRRELS

15. PRAIRIE DOGS

16. GRAY SQUIRREL

17. RED SQUIRREL

18. FLYING SQUIRRELS

19. POCKET GOPHERS

20. POCKET MICE

21. KANGAROO RATS

22. AMERICAN BEAVER

23. RICE AND COTTON RATS

24. HARVEST MICE

25. WHITE-FOOTED MICE

26. WOODRATS

27. NORWAY RAT

28. HOUSE MOUSE

29. MEADOW VOLE AND OTHERS

30. COMMON MUSKRAT

31. JUMPING MICE

32. NORTH AMERICAN PORCUPINE

33. AGOUTIS

34. PACA

Family Canidae: Dogs, Foxes, and Wolves 153

35. DOMESTIC DOG

36. COYOTE

37. GRAY WOLF

38. ARCTIC FOX

39. KIT AND SWIFT FOXES

40. RED FOX

41. COMMON GRAY FOX

42. BLACK BEAR   43. BROWN (GRIZZLY) BEAR   44. POLAR BEAR

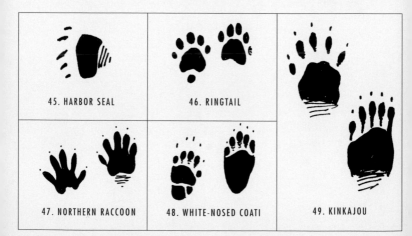

45. HARBOR SEAL   46. RINGTAIL

47. NORTHERN RACCOON   48. WHITE-NOSED COATI   49. KINKAJOU

50. AMERICAN MARTEN AND FISHER

51. WEASELS

52. AMERICAN MINK

53. TAYRA

54. WOLVERINE

55. AMERICAN BADGER

56. NORTHERN RIVER OTTER

57. SEA OTTER

58. SPOTTED SKUNKS

59. STRIPED AND HOODED SKUNKS

60. HOG-NOSED SKUNK

61. DOMESTIC CAT

64. JAGUARUNDI

65. CANADA LYNX

62. COUGAR OR MOUNTAIN LION

63. OCELOT

66. BOBCAT

67. JAGUAR

68. HORSE AND BURRO

69. WILD BOAR

70. COLLARED PECCARY

71. ELK

72. DEER

73. MOOSE

74. CARIBOU AND REINDEER

75. PRONGHORN

76. AMERICAN BISON

| 77. MOUNTAIN GOAT | 78. MUSKOX | 79. DALL'S AND BIGHORN SHEEP | 80. BAIRD'S TAPIR |

| 81. BIRDS | 82. AMPHIBIANS AND REPTILES | 83. INSECTS AND OTHER INVERTEBRATES |

A FIELD GUIDE TO

ANIMAL TRACKS

INTRODUCTION
WHAT HAS HAPPENED HERE?

WHO LIVES IN the forest? What creatures inhabit the banks of streams, the shores of lakes, or the sands of the desert? What are the animals that leave footprints in the mud and trails in the snow? What has gnawed the bark or clipped the twig?

Hunting is about as old as life, and tracking is an ancient science, but it would be difficult to determine when land animals began to notice the tracks left by their companions in the mud or sand. It is not likely that the amphibians of prehistory were trackwise. What of the reptiles of the Permian Period, and those awkward reptile-mammal creatures that were the forerunners of the graceful present-day mammals? Certainly many mammals are trackers, and they may be the first to have learned the art. But they track by scent. Although a dog assuredly sees the rabbit track, and recognizes it as a track, he can hardly sense that the shape, size, and arrangement spell "rabbit." However, the first whiff with his nose tells him what it is.

I was traveling one day with a cougar hound in a light snow in the deep forest of the Pacific Coast. We found a wolf trail and I was interested in noting the size of the tracks and the stride. The hound poked his nose deep into the footprints to learn their identity. Then as we followed a little way I noticed the dog sniffing curiously at various twigs overhanging from either side. The wolf had merely brushed them in passing, but it had left a record sufficient for this master tracker who was with me. What need had he for a knowledge of size or shape of footprints?

Humans have lost the power of tracking by scent but have developed greater intellectual refinement. We can read a more complex story in footprints than mere identification and direction of travel. The Indians and Eskimos and the experienced woodsmen have learned to read imperfect track records with remarkable skill, just as a trained ornithologist recognizes a bird without seeing the details of its plumage. The more recent use of tracking ability is in the field of natural history.

It has become a popular pastime to go looking for birds, but what a field there is open to us in becoming familiar with mammals! They are furtive, usually silent, and very often go about at night. You don't find them readily with field glasses, and you don't total up a long list on a "mammal walk," as you do with birds. Nevertheless, their very aloofness is the challenge, and your sleuthing instinct is aroused. Once alerted to this fascinating game, you can never again pass the muddy margin of a stream without instinctively looking to see what has passed by there. You will speculate about every trail in the snow, big or small. You will find that a complex mammal world, which you never suspected, has opened up for you. Driving along a desert road, you can stop and by a five-minute walk in the sand gain some acquaintance with the kangaroo rat, kit fox, or black-tailed jackrabbit through the medium of their footprints.

Perhaps it is best for us not to specialize too much. Bird enthusiasts can add to their enjoyment and understanding by some interest and skill in reading the record of mammals. Hunters and researchers will better understand the ecology of their quarry if they recognize the signs of all animals. Naturalists go forth to enjoy what they can find, be it bird, mammal, insect, plant, or the music of a mountain stream.

On a wintry day in Wyoming as I was traveling along the foothills in open country, with aspen groves on the hills, dark evergreen forests above, and the Teton Mountains across the valley, I came upon the tracks of a jackrabbit. Obviously the animal had been in a desperate hurry. How desperate was soon revealed, for the rabbit track was joined by the track of a coyote. The rabbit had dodged deftly and fast. The coyote swerved too, but had slid in a wider turn. Again the rabbit turned, and the coyote swung in close pursuit. Time after time this was repeated. The rabbit was doing well. Could he gain enough distance to escape by a speedy straightaway?

Suddenly there was a third track. A second coyote had been cruising along a little farther up the hill. His trail of long leaps led diagonally down, just in time to close in when the jackrabbit made one of his desperate uphill turns. A few drops of blood here, and the two coyote trails led off together.

One fall I was looking for elk grazing areas in the upper Yellowstone country. The question was, did the elk summer here, and if so, how extensively? I entered a high basin with open slopes. There were the elk tracks, and there were their droppings. The character of the dung—more or less flattened or formless—revealed that this was summer range, that the animals had found succulent green forage. Winter or autumn droppings would have been hard and pelletlike. Such indications help to build up the facts in the life story of an animal and are useful to the scientist as well as the amateur observer.

Another time, driving along a road in Utah, I glanced at a rocky outcrop with low cliffs. It was possible to know in that glance that woodrats lived in those rocks. Why? Because there were the tell-tale white spots, visible at a distance, where the little animals had left their excretions. Farther along I could see the location of a nest on a high cliff. This proved to be a raven's home. It had been revealed by long vertical streaks of "whitewash," the smears of guano that accumulate at such nests. They are distinctly different from the marks left by woodrats.

One day in the Wichita Mountains of Oklahoma I watched a flock of sandhill cranes alighting at the marshy shore of a pond. After they had left I looked over the ground. I was struck by certain similarities between the tracks of these cranes and those of the wild turkeys that I had seen nearby. I obtained plaster casts of both for more leisurely study. The casts clearly revealed the trim feet of the crane, with a finer "fingerprint" pattern on the undersurface of the toes, in contrast to the coarser aspect of the turkey track.

In April 1949 a friend and I were clambering through the mountains of southern New Zealand, looking for signs of red deer and wapiti. We came on a narrow trail in the moss and litter of the forest floor, leading off from the base of a huge tropical beech. In America it would have been noted as a squirrel trail, but there are no squirrels in New Zealand. We were thrilled when finally it be-

A brown bear's tracks follow a swollen glacial river along the Alaska coast.

came clear to us that here was the track of the kakapo, a flightless parrot now becoming very rare. Unfortunately we didn't see the bird itself, since it is nocturnal in habit like the owl.

Familiarity with animal signs—tracks, droppings, gnawings, scratchings, rubbings, dams, nests, burrows—can open up a delightful field for the outdoor traveler. The mountain climber and the touring skier can make interesting observations in far and high places, increasing the pleasure of their excursions afield. The hiker, the visitor to national parks, the wilderness traveler in national forests, can add much to their experience by sleuthing about for traces of the unseen inhabitants of the region.

If you are vacationing in Big Bend National Park you will be interested in a possible bobcat track in the mud beside the Rio Grande, or the mountain lion "scratching" in the trail in the Chisos Mountains, or the nibblings on a prickly pear on the desert flatlands. If you spend some days in Denali National Park of Alaska, you have a chance to see at first hand the peculiar rounded track of the caribou, the hay piles of the singing vole (or "hay mouse"), or the pellets of the gyrfalcon at its aerie in the cliffs. The Boy Scout can enrich his excursions with an interest in animal signs. The serious field biologist requires facility in reading the story left by the passage of mammals. Whatever the purpose, these little stories in wild country can be the means of great esthetic adventuring.

The coyote can read the news of his territory with his nose; and what a wonderfully sniffy time your dog has as he follows you through the woods. It is the purpose of this Field Guide to give some aid to humans in the interpretation of animal "sign," not in the man-

The typical trail of a walking bobcat in shallow substrate, an overstep walk, in western Massachusetts.

ner of the coyote or dog, but by conscious evaluation of what we see before us. Its object is to encourage the study and enjoyment of the outdoors and the wildlife in it.

Mammals are not the only ones to leave signs of their presence. Accordingly, I have included a few tracks and other signs of representative birds, reptiles, amphibians, and insects. Even the weather leaves a few signs, such as lightning marks, shown in figure 197, b. The main concern, however, is the mammal world of North America, from Mexico to the Arctic.

READING THE SIGN. Reading tracks is not easy. Just as a detective, with certain broad principles in mind, finds each situation somewhat different, so the animal tracker must be prepared to use ingenuity to interpret what he or she sees. A track in the mud may look different from one in dust, or in snow, even if the same individual animal made them. A track in snow is different after a warm sun has shone on it—enlarging and distorting it. An *average* or standardized track drawing in a book may not look like the one you are trying to identify. The fact is that in many instances the track you find may not seem to fit anything because parts are missing; it may not show all the toes, it may be off-shape because of irregularity of the ground. Then there are the variations due to age of the animal, or the sex. Generally speaking, the front track is likely to be different from the hind track of the same animal, with a different number of toes, different shape and size. And perfect tracks are not always found.

The same applies to droppings. Those of a half-grown coyote may resemble those of a full-grown fox. If a fox or coyote has been feeding on bulky food, with a lot of fur or feathers, the dropping is likely to be unusually large. If the same animal has been eating lean rations, the dropping may be very small. Scats composed of fur also look radically different from those composed of fruits or vegetation.

Therefore, use the material in this book as a guide rather than a rigid key. In the consideration of each species, all possible suggestions will be given to help unravel the story of animal signs. Space does not permit illustration of all the variations you will find—and in some cases it has been impossible to obtain any tracks. The policy has been adopted to illustrate, whenever possible, actual tracks, with place and date and character of the snow or ground, and to show several variations. It has been possible to rely mostly on personal experience for the contents of this guide, but no one writer can claim complete information, and there are gaps in the story which will be pointed out in appropriate places. In some cases I have drawn on the information and experience of others, as indicated in the text.

Now a word of encouragement to the beginner. After you have entered this field of mammal study, you will gradually become conscious of a *knack* in reading signs. In familiar territory you will become adept at interpretation of fragmentary evidence and it will not be necessary to have it spelled out in detail. However, even professional naturalists, if they are honest, will freely say that sometimes they are stumped by what they find.

For the experienced student, I can only hope that this handbook will be helpful. He or she will recognize the interdependence of those of us who work in this field. I would here express my appreciation of published records, which have been consulted and duly listed elsewhere, and the enthusiastic assistance of coworkers. With humility, and with full knowledge of how great and complex is this field, I offer this guide on animal signs.

PRESERVING TRACKS. To the field biologist and environmental educator there may be an advantage in preserving for future study a clear record of tracks in tangible form. To the student or wilderness traveler who might make a hobby of tracking it is also worth while to "take home" souvenir tracks, though this is sometimes awkward. I was confronted by an incredulous customs inspector on my return from Panama. He could hardly believe my sanity when I told him the large crate among my belongings contained "animal tracks."

If one is accustomed to sketching and carrying a pocket tape measure, a good way is to make careful drawings of the tracks, either natural size by the aid of the tape measure (or measured by the pencil) or drawn to scale. One should take note of the lengths of a few sample strides or leaps, as well as the dimensions of individual prints.

Under rare circumstances it is possible to cut around the track with your knife and lift it out intact. I found it possible to do this in the mud of a dried lake bottom in Nevada, where pronghorn and coyotes had left footprints when the mud was just right for leaving a firm, sharp print.

Generally, however, for "take home" tracks I have relied on plaster of Paris. I have rarely gone on a field trip during the past twenty-five years without a supply of plaster. The result is an extensive collection of track casts. These, with numerous field drawings, form the basis for the illustrations in this volume.

The method is simple. First, carry your plaster in some tightly stoppered receptacle—a widemouth bottle or jar, or friction-top can, or if in a paper sack, nest two or three sacks inside each other to form several thicknesses of paper. Loose plaster plays havoc if it gets into your packsack.

Upon finding a track in mud, I usually take a tin can, put in a little water, add some plaster, and stir it quickly with my fingers or a stick until the mixture forms a thin batter; or I may put the plaster in first, which is probably a little better. I estimate the amount of "batter" needed for the track or tracks before me. If the mixture appears too thick, I hastily dip in a little more water with my hand—*not too much*—until it seems right. It is important of course to add water before the mass begins to stiffen, and to work fairly fast if the first mixture begins to come out thick. A very thick plaster paste begins to set almost immediately. A thinner mixture gives you more time. When the plaster has been poured into the track I wait patiently or busy myself with other things for twenty minutes or more to allow it to set hard. It is not wise to attempt to pick up the cast too soon. You can test a cast to see if it is ready to be removed by running a finger along its exposed side. If ANY white appears on your finger, be patient and leave it longer. Sometimes, if I am on the "out trail," I leave the casts and pick them up on my way back.

If the plaster is too thick so that it does not flow readily, it will begin to set while being poured and will not enter all the crevices of the track as it should, perhaps failing to catch sharply enough the details of feet like those of the porcupine or bear. If too thin or watery, it will run all over the place, taking forever to harden and acting queerly in many ways. Sway the solution in the can as you mix it; when it loses the watery quality and begins to move a little sluggishly, with the first sign of "reluctance," it is about ready to pour. Plaster of this consistency, like a thin batter that pours fairly easily, will make a hard cast. When too thick, or if diluted after it has begun to set, it will harden into a chalklike consistency that scratches or abrades easily.

You may wish to include in the cast several tracks, the whole four-foot pattern of a squirrel or weasel or skunk, or you may want to include two sets of four to show the stride or jump. In that case pour plaster into one set of tracks, then trail it across the ground in a strip to the other set. With a long cast of this kind I often lay one or more sticks or pieces of wire into it for reinforcement while the plaster is still soft.

Sometimes when I am short of plaster and want to stretch it, or when I am greedy and want to get "one more track" with the batch I have prepared, I pour in enough to cover the bottom to make sure of the details, then put in sand or stones or sticks and add the remaining mixture on top. This will do in an emergency. If you find that you have underestimated the amount needed, more can be mixed and poured on top of the plaster already there, in order to make sure the cast will be thick enough for safe handling.

You should be sure that the plaster overflows the margins of the track to make certain you have all of it, with all its edges and claw marks. If the track is on sloping ground you may have to place a dirt ridge, or stones or sticks, on the lower side to keep the plaster from running away. The proper method is to place a "collar" of heavy paper or other suitable material around the track in order to retain the plaster within bounds and make a *thick* cast that won't break. I confess that, with the urgency of other work, I have seldom bothered with this detail.

To pick up the cast *after it is hard,* cut around it with a knife, gouge out some dirt from under the edges all around, then lift out the piece from a point well underneath. Take it to water and wash off the mud, using a toothbrush with care, if you have an old one at hand for such purposes. Should enough water for washing not be available, simply brush off the surplus mud and wash the cast later. Caution should be taken while washing the cast, for if you clean your cast completely of all earth, the tracks lose contrast and become very difficult to see. A thin layer of native soil is what is needed to create the relief necessary to see every detail of the footprint.

These are crude and rough field methods. The serious student may wish to work out greater refinement. Some naturalists employ improved techniques. Dr. E. Laurence Palmer of Cornell University, for example, suggested that to delay the hardening of plaster you might put a little vinegar in the solution.

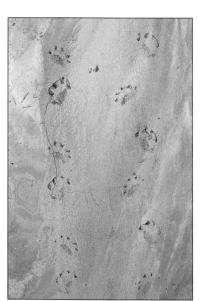

If the track is in dust great care must be taken. My own method is to pour the solution directly into the track, but from a point as close above the track as possible. It should not be dropped in with any force.

The beautiful trail of an unknown turtle (likely painted turtle) in southern New Hampshire.

Some naturalists use paraffin or a candle in place of plaster of Paris. It is admittedly simpler to carry some candles on a field trip. Vernon Bailey recommended dripping a lighted candle into the track. But one difficulty I have found is that unless there is an ample quantity of melted paraffin or wax to pour in rather quickly, you get an irregular surface that does not give continuous detail, since the paraffin tends to harden while pouring.

Tracks in snow are difficult to cast, and obviously a candle cannot be used. Even a fairly cold plaster solution tends to melt the bottom of the track, and it is disconcerting to find that the batter has poured right on through. To avoid this the mixture should be made as cold as possible by stirring some snow in while mixing. I have also tried dusting in a fine layer of dry plaster first, as a base, allowing it to form a slight lining in the track. Dr. Palmer, mentioned above, devised an excellent method for cold weather. With an ordinary atomizer containing very cold water, he sprayed the surface of the track enough to form a thin coating of ice that would hold the cold plaster of Paris mixture. Modern trackers often employ Snow Wax Print (Kinderprint Co., 800-227-6020), which they first spray liberally into the track to hold its form, and then follow up with plaster.

I have experimented with a substance carried by hardware stores called water putty. It is a yellowish brown powder to be mixed with water into a thick paste for patching cracks and holes in cement, plaster, and so on. I find that by mixing it thinner, pretty much as with plaster of Paris, it makes excellent casts. Another advantage is that by mixing in 50 percent of fine sand, or almost any soil at hand, I have obtained a very hard cast. Perhaps 25 percent of sand is safer to experiment with, though in an emergency, when I was short of material, I have put in much more than 50 percent of sand. However, the more sand, the longer it takes to harden. If you attempt to lift out this sand mixture too soon, the cast may break up or crumble. Leave it a long time.

Although this is a material not yet fully proved, some of the casts I have made, especially those with a moderate sand mixture, seem very hard and durable and may prove to be less destructible than plaster casts. This is a field for experimentation; but I would offer a word of caution: for small tracks with fine detail, use water putty pure; large coarse tracks are suitable for the somewhat grainy surface you get by adding sand.

PRESERVING SCATS. The field biologist may also wish to make a collection of scats as reference specimens, for use in studies of food habits, or for any other purpose requiring identification of such material. The first essential is to ensure proper identification in the field, preferably by tracks where the scat is found, or by collection in a locality where it is certain that only one kind of mammal of the type involved is present. For instance, it is helpful if a coyote scat is obtained where it is known that bobcats are absent (though in some parts of the country bobcat scats are fairly distinctive).

The scats, or droppings, if desired as permanent specimens, should be kept in their original shape by suitable packing in the field. In the case of specimens containing animal remains, such as bones and hair, moths or dermestid beetles will destroy them just as they destroy museum skins. I have found it helpful to coat such scats liberally with shellac or glue varnish, using a brush, or with any spray shellac, allowing some of the application to soak into the substance. I have not found it necessary to treat material consisting of plant remains in this way, except to ensure holding the shape.

Be warned that there are inherent dangers in handling some scats, particularly those of certain rodents and raccoons. In parts of North America, hantavirus is carried in rodent scats and can be inhaled in confined places where droppings accumulate. The roundworm *Baylisascaris procyonis* can be found in some raccoon scats, and if inhaled could be fatal. Leptospires, found in mammal urine, are also dangerous to humans. These are not reasons to avoid scats completely. I dissect scats very regularly and follow simple rules: dissect raccoon scats with sticks, and avoid enclosed areas with heavy accumulations of rodent scats.

On several occasions I have found droppings of elk, deer, and mountain sheep formed of pure clay, in the vicinity of mud or clay licks. These make excellent specimens, being perfectly hard and of finely modeled form. Occasionally, too, I have found marmot droppings on arid buttes that appeared to be coated with reddish clay—possibly windblown.

Of course there are only special types of scats that can be collected, but a series of typical, well-identified material of this kind, including certain bird droppings and castings, can be a useful aid in certain field investigations and schoolroom projects.

It occurs to me that a scat collection of extant animals, properly identified, should be of importance to the vertebrate paleontologist, who at times has to deal with coprolites found in the fossil record. For such a purpose, particular attention is directed to the difficulty of identification in certain animal groups, a fact emphasized in the following pages.

STAGE SETTING FOR TRACKS. Many years ago, in his *Two Little Savages,* Ernest Thompson Seton described a method of getting good tracks by preparing for them. If you know of a place where mammals habitually travel, as on a trail, or near a den tree, or any other conceivable situation, you may prepare a smooth area of dust or soft mud, as the case may be, over which a traveling mam-

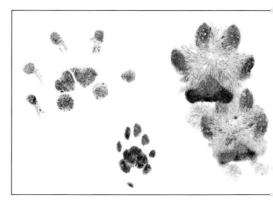

Two-dimensional soot tracks on contact paper: opossum in the upper left, ringtail in the lower middle, and on the right, gray fox; all captured in California.

mal will leave good tracks under controlled conditions. It may be possible to use bait to attract certain species over prepared ground. Old tracks may be erased by smoothing over the surface like a clean slate. This general idea has possibilities for expansion and the exercise of ingenuity, not only for the casual student but also for the field biologist who wishes to delve into the field of population studies and ecology.

CATCHING TRACKS. Sooted track plates are now widely used in wildlife research, allowing one to "catch" and keep the tracks of wild beasts. This can be done in one of two ways. You can place a completely sooted and baited track plate in an area of high traffic, and then read tracks directly from the plate. (Tracks will appear as negative space, the area where soot has been removed.) Or sooted plates can be used in conjunction with white contact paper, which allows you to keep a permanent record of the animal. The second method is only slightly more complicated, and the resulting track collection is certainly worth the effort. Understand that the animal must first step on the soot, which sticks to the foot, and then onto the sticky side of the contact paper, where the track will be recorded.

Use thin sheets of aluminum as the base. Black soot may be beautifully applied with the sooty flames of acetylene torches, or a small container of diesel fuel (with a wick). A garden tiki torch can also be used in a pinch. I have also heard that glass can be sooted by running sheets over candle flames, and the results are decent.

It may take some adjustment to confidently read the two-dimensional tracks on contact paper, as they look quite different from tracks made in natural surfaces. Experiment with different setups and make sure to include some protective cover if your area receives regular rainfall. For further details on track plates, refer to USDA General Technical Report PSW-GTR-157, called "American Marten, Fisher, Lynx, and Wolverine: Survey Methods for Their Detection."

CAPTIVE ANIMALS. Many people find it useful or interesting to keep small mammals in captivity for a short time for close observation, for sketching, or for serious study of mannerisms or other reactions throwing light on their life histories. This subject has many ramifications and cannot be treated fully here. But the reader will find in the bibliography a few references to articles on live-trapping and types of traps used.

In this connection I would make a few comments. During chilly or cold weather a mouse sitting all night in a live trap without exercise becomes thoroughly chilled, and in really cold weather may die before morning. If a little cotton is placed in the trap chamber the mouse will make a nest and be comfortable. Also, there should be enough bait to serve as food during the night.

When small rodents are being transported, either in a car or by express, the container should have the usual food, and in lieu of water, some form of succulence, such as an apple or potato. It is well to nearly fill the container with excelsior or similar stiff but resilient material, so that the rodents may form burrows and not be injured by heavy food substances rattling about.

LITERATURE. A rather random list of titles has been selected from the literature on mammals and placed at the back of the book. This bibliography may at first glance appear formidable to the beginner. I would suggest that the scientific literature on mammal life history or ecology is not unduly technical, however, and does contain fascinating information on the lives of these creatures that is not suspected by the general public. To the field biologist these titles are familiar, but they may be useful as reminders.

For all readers such literature, if sought from time to time, would help outline the broad field of mammals and increase the enjoyment and understanding of the less well known animal life of our country.

SPECIES
ACCOUNTS

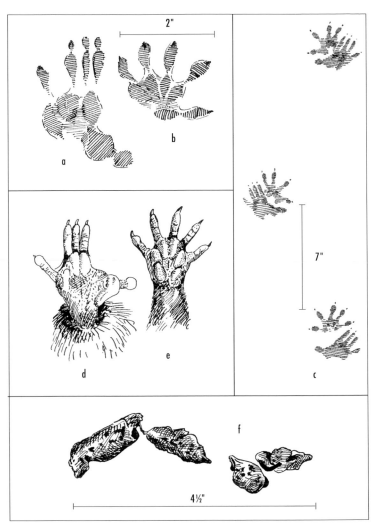

Figure 1. Virginia Opossum

a and b. Hind and front tracks, respectively, in mud (Oklahoma).

c. Typical walking pattern of opossum, in mud (Washington).

d and e. Hind and front feet, respectively, showing unusual arrangement of the toes on the hind foot.

f. Opossum scat.

FAMILY DIDELPHIDAE: OPOSSUMS

THE VIRGINIA OPOSSUM, *Didelphis virginiana*, is an inhabitant of much of the eastern United States, as well as the western seaboard as far north as Washington. Undoubtedly it will eventually occupy the greater part of the country except the coldest and the most arid regions, for which it is not adapted. In the far north it is not unusual to find opossums that have lost toes and tail to horrible frostbite.

This slow-moving animal is not particular about what it eats. Small mammals, birds, eggs, insects, fruit, carrion, and garbage are all acceptable. Usually found in wooded areas, swamps, along streams or lakeshores, it seeks its shelter in old dens of other animals, in crevices under rocks, in hollow trees or logs, and often seeks safety by climbing a tree. In Texas I found the opossum in rough terrain, in oak forests, and in southern California in arid brushland.

Front and hind tracks of an opossum along a riverbank in Washington State.

I first found the strange, distinctive opossum tracks (shown in figure 1, a, b, c) years ago in Oklahoma, in the mud near a pond. Note the shape of the peculiar hind foot, and that the "big toe," which is slanted inward or even backward, does not have a claw. There are five toes on each hind foot, but the three middle toes tend to remain in a close group, separate from the others. Front tracks measure 1–2 5/16 inches long and 1 1/4–2 1/2 inches wide. Hind tracks are 1 3/16–2 3/4 inches long and 1 1/2–3 inches wide.

An uncommon walking pattern is similar to that of the raccoon —front and hind tracks side by side. More often the hind foot falls on top of or a little posterior to the front track (c), and the steps may vary from 4 to 9 inches. When moving very slowly, the opossum may drag its tail, but while it walks normally or trots about, the tail is held aloft.

Opossum scats are unfortunately not distinctive and will vary in accordance with the kind of food eaten. Figure 1, f, gives their general character. They are also difficult to locate in the field. I myself have found them only when removing unfortunate opossums from live traps I'd set for forest carnivores in southern Mexico and along well-used runs protected under bridges in southern California.

So far as we know, the opossum has no call beyond a low growl or a slight hissing when disturbed. At any rate it gives out no voice that would help direct attention to it. In fact a behavior for which the opossum has gained renown is playing dead. In the face of predators, it is often the strategy of this animal to roll over, feign death, and release an unpleasant acrid odor rather than to escape by speed and evasion.

FAMILY SORICIDAE: SHREWS

THE TINY SHREWS, often mistakenly referred to as "those sharp-nosed little mice," or "moles," are active, high-spirited creatures that keep busy hunting insects and other small animal life, or eating carrion. A large variety of species and subspecies is included in the genus *Sorex*, the long-tailed shrews, and these range over most of the North American continent. I have found the Arctic shrew, *Sorex arcticus,* of the Far North particularly interesting, a handsome morsel of animal life, rich brown above, paler on the sides, white below, in an attractive tricolor pattern. In the earlier days of travel by dog team, they used to find our stored feed, dried salmon, and nibble at it persistently. On one occasion on the Alaska Peninsula, I heard thin squeaky voices in the tall grass. I watched quietly and had glimpses of tiny forms flashing by the little openings in the dense cover. I could only guess what might be the intimate affairs of those diminutive mammals.

Shrews are chiefly nocturnal, but may be found about in daytime as well. These so-called long-tailed shrews use the tun-

Cinereus shrew tracks along a riverbank in central Alaska.

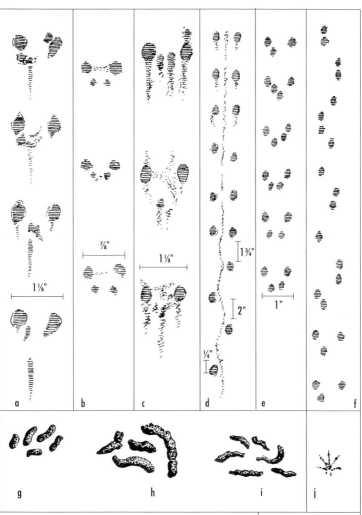

Figure 2. Shrew tracks and droppings

nels of moles, mice, and the larger short-tailed shrews, and they find shelter under bark, logs, and the forest litter. On the Pacific Coast I have found burrows that appeared to have been dug by the shrews themselves. Their nests are globular, of shredded leaves or other material, even rabbit hair.

Shrew tracks are most evident in snow, though they may be looked for in areas of dust or mud, especially under logs and boulders where soil is protected. In snow the shrew trail is much like that of the white-footed mouse, but, at least in the case of the small long-tailed shrews, the tracks and track pattern are smaller. The straddle of the shrew trail will measure between ¾ and 1 inch, and occasionally as much as 1 ⁵⁄₁₆ inches, as compared with 1 ¼–1 ¾ inches for white-footed mice. Also, the shrew track pattern appears a little shorter, as if the hind feet do not reach as far beyond the forefoot tracks as do those of the mouse. We must note, too, the depth and texture of snow. In loose fluffy snow the mark of the tail may show, as well as drag marks of the feet. On a thin film of snow over a snow crust or on ice, tail marks are normally lacking and footprints more distinct.

The gait of the shrew varies, even in a section of some 10 or 12 feet of its trail. Notice the many variations shown in figure 2.

In soft snow the shrew has a habit shared by several other mammals. It will dive in and travel under the surface. The course of such a snow tunnel often may be detected by the slight ridging

Figure 2 (opposite)

a. Bounding trail in snow, showing trail drag: a through f show tracks from Wyoming; and all trails are moving up the page.

b. Another bounding trail in light snow, without tail marks.

c. In this snow trail the feet were dragging.

d. Here the tail made a continuous mark as the animal switched from a trot into a bound.

e. On firm snow the animal didn't flounder; the trail shows common variations.

f. More irregular pattern here, often found in a short length of trail when the animal slows to a trot or walk.

g. Scat of *Sorex* sp., about natural size (Wyoming).

h. Scat of northern short-tailed shrew, *Blarina brevicauda,* natural size (New York, furnished by W. J. Hamilton Jr.).

i. Scat of *Cryptotis parva,* natural size (furnished by E. P. Walker).

j. Front track of *C. parva,* natural size.

k. Hind and front tracks of *B. brevicauda* (New Hampshire).

of the snow. The veteran naturalist Dr. E. W. Nelson reported an amazing incident of following such an under-snow shrew trail on the Yukon River for a distance of a mile or more.

While resting beside a tiny stream in an open mountain meadow in Wyoming, I saw a shrew swimming under water, upstream, encased in a silvery film of air. This was the large American water shrew, *Sorex palustris,* the one with the pure black and silvery white uniform. It is especially adapted to life around borders of streams and ponds and takes readily to water. I have never seen its tracks, but they would be larger than those of its smaller relatives, about the size of those of the white-footed mouse.

Blarina brevicauda, the northern short-tailed shrew, represents another group of large size. I found these animals commonly in the plowed fields, and many other locations, in Minnesota and eastern North Dakota. In winter they shared with the meadow voles, or field mice, the space under the shore ice shelf along the Red River, formed when the water had dropped to a new ice level. They have runways in moss and other vegetation after the manner of the field mice, make their own burrows, and also use the burrows and runways of small rodents. These are common throughout the eastern half of the United States and adjacent parts of Canada.

The track patterns of short-tailed shrews are more similar to those of voles than to smaller shrews. Their preferred gait is a trot, leaving a trail like a miniature coyote, rather than the bounding patterns so common in mice.

Track comparisons:

Sorex cinereus—front, ³⁄₁₆–¼ in. long, ³⁄₁₆–¼ in. wide; *hind,*
³⁄₁₆–³⁄₈ in. long, ³⁄₁₆–¼ in. wide

Blarina brevicauda—front, ¼–³⁄₈ in. long, ⁷⁄₃₂–1 ¹⁄₃₂ in. wide;
hind, ¼–³⁄₈ in. long, ³⁄₁₆–1 ¹⁄₃₂ in. wide

There are some small species of shrew, slightly over 3 inches long, including the tail. These are the pygmy shrew, *S. hoyi,* and the least shrew, *Cryptotis parva.* The least shrew occupies the eastern half of the United States and becomes very common in Mexico and Central America. The pygmy shrew is more northern in its distribution, being found in the northeastern United States and across the continent from Labrador to Alaska. One should also mention the rare Crawford's desert shrew, *Notiosorex crawfordi.* It is found only among the desert shrubs of Texas, New Mexico, Arizona, southern Nevada, and California, and down into Baja California and Mexico proper. Probably these members of the shrew family are the smallest of all mammals, and figure 2, j, could very well represent the smallest mammal track in the world.

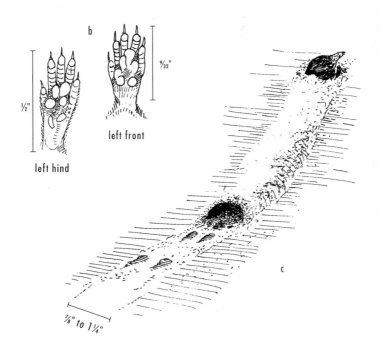

Figure 3. Arctic shrew

a. Arctic shrew.

b. Feet, enlarged to show the tubercle pattern of the underside.

c. The shrew dives into the snow, to come out some distance beyond.

FAMILY TALPIDAE: MOLES

THERE IS AN amazing confusion about moles and certain rodents. Pocket gophers, and in some places even field mice, are often referred to as moles. Not everything that works underground is a mole! Therefore it is worth while to give some attention to the groups of moles and the signs they leave for identification.

Moles have soft fur, a compact body with a rather naked snout, minute eyes, and spadelike front feet, altogether well adapted to their subterranean life. Footprints are rare, but earth mounds and tunnels reveal their presence easily enough (figures 4 and 5). Front tracks of the hairy-tailed mole measure ⅜–⁹⁄₁₆ inch long. Hind tracks are ⅜–½ inch long and ¼–⁷⁄₁₆ inch wide.

"Temporary" runways consist of raised ridges in the surface of the ground, pushed up as the mole progresses just under the surface. These will vary in appearance with the character of the soil. A permanent network of tunnels lies deeper underground. They must not be confused with the earth cores left in snow tunnels by pocket go-

A beautiful ridge created by a mole tunneling along the surface of a riverbank in western Massachusetts.

The trail of an unknown mole species in western Massachusetts. The sandy environment suggests that it is a hairy-tailed mole.

phers (for which see figure 43, b). Mole ridges have a tunnel underneath; earth cores left by pocket gophers, lying on top of the ground after the snow melts, are solid. In certain types of firm earth there will be tracks in the crust where the mole has raised these ridges. In at least one recorded instance such solid earthen casts were found in an area occupied only by the Townsend's mole, *Scapanus townsendii*, pocket gophers being absent. This showed that occasionally the mole too may excavate into snow tunnels. There is also an insect that makes a tunnel like the mole's, but very much smaller; from this it is called the mole cricket (see figure 189, a).

In excavating, the mole pushes the dirt out to let it roll where it will. Consequently a molehill has the appearance of an eruption, with no indication of burrow entrance. The pocket gopher pushes the dirt away mostly in one direction, so that the entrance is at or near one edge of the mound, at least off center, and the final earth plug marking the entrance, in a depression, is generally obvious.

The brownish black scats, or droppings, of the hairy-tailed mole are between ⅜ and 1 inch long, and ³⁄₃₂–¹⁄₁₆ inch in diameter, and are somewhat cylindrical, tapering to points. They dry quickly to a hard consistency with finely pitted surface, composed of soil particles and chitinous remains of insects, which appear as shiny spots. Look for great accumulations of their scats in latrines formed at burrow entrances under logs and other such cover.

The eastern mole, *Scalopus aquaticus*, of the East and as far west as southern Minnesota, Nebraska, Oklahoma, and Texas; the broad-footed mole, *Scapanus latimanus*, of the Pacific Coast states; and the hairy-tailed mole, *Parascalops breweri*, of the northeastern states and adjacent parts of Canada, are similar in appearance and similar in habits. The mound and runway structure de-

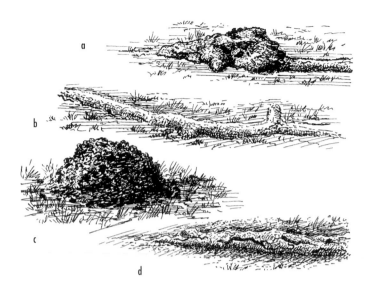

Figure 4. Mole sign

a. A ridge from the right ending in a small mound, which is 5–8½ in. in diameter and 2 in. high. The ridge at the right is 1½ in. wide (Point Lobos, California).

b. Mole ridge at Point Lobos, diam. 2 in. In loose sand the diameter varied from 5 to 6 in.

c. Typical molehill.

d. Another type of ridge, where the firm crust was raised and cracked as the animal passed underneath (near Redding, California).

e. Front and hind tracks of a hairy-tailed mole in wet sand (Massachusetts).

f. Trail of a hairy-tailed mole in wet sand (Massachusetts).

front

hind

e

f

a

b

Figure 5. Molehills in Wisconsin
a. A line of molehills on a golf course.
b. Typical lumpy structure of molehill.

scribed above apply pretty well to all of these, though the mounds of *Parascalops* are smaller and the runways less marked than those of the other two. However, there is great individual variation.

The star-nosed mole, *Condylura cristata,* also of the northeastern part of the United States and adjacent parts of Canada, differs somewhat in habits, and may come out of subterranean runways and continue them in the grass, or through the snow, and may be found on the snow surface. It also enters water through underwater entrances, and travels under the shore ice of stream borders.

The little American shrew mole, *Neurotrichus gibbsii,* 4½ inches long, of the Pacific Coast from southwestern British Columbia to Monterey County of California, is even less orthodox. Not only is it a good swimmer but it may climb into bushes. Its runs, about ⅝ inch wide, are often quite conspicuous within the debris layer, rather than in the earth as described with other mole species.

Lloyd Ingles speaks of it thus:

> Among the litter of rotting logs and dead leaves in the shady ravines close to the coast of California are little runways that form an irregular but intricate network of semisubterranean tunnels and passages... If the shade is not too thick, one may possibly catch a glimpse of a tiny mammal, resembling a large shrew, tapping the ground with its long snout, as it walks slowly along in search of food. This is the shrew-mole, which . . . resembles both a shrew and a mole, and is about intermediate between them in size.

This is one of the rare finds for the enterprising naturalist.

ORDER CHIROPTERA: BATS

No NEED TO LOOK for the footprints of bats! Their limbs are so enmeshed in a highly developed flight membrane that they do not bother to walk, except for a short distance when they find themselves on a horizontal surface, or when they crawl into hiding in a narrow crevice to roost. But they can walk, and I have seen one swim ashore when it accidentally got into the water of a pond. The sketch in figure 6, a, shows the trail of a long-eared myotis, *Myotis evotis*, that was turned loose on fine sand. Note the outside prints of the front wrist, just *outside* the print of the hind foot, and the scrape of the wings at the end when it flew off.

Bats may often be detected by the deposit of guano on the ground or floor beneath their roosting places, or by the stained entrances to such roosts in trees or walls. Figure 6 illustrates bat droppings, which may be distinguished from those of mice by the insect fragments of which they consist. There is also a tendency to be segmented, as shown in the enlarged view.

The stained wall denotes the entrance to a bat roost in an abandoned shack in Santa Barbara County, California.

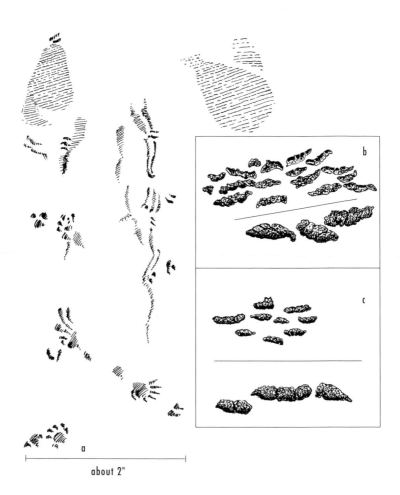

Figure 6. Bat tracks and droppings

a. Tracks of a bat walking on fine sand, and taking off in flight.

b. Guano, bat of unknown species, probably *Myotis* sp. Upper samples, natural size; lower, enlarged (Jackson Lake, Wyoming).

c. Droppings of long-eared myotis, *M. evotis*. Upper samples, natural size; lower, enlarged.

about 2"

Looking for bats can be alluring. Some, such as the eastern red bat, *Lasiurus borealis,* and the large hoary bat, *L. cinereus,* may be found during the day hanging from the twigs of trees. Others hide in crevices behind the bark of trees, under loose shingles, or in cracks in large rocks. It is well known that several species swarm in caves and assemble in attics and barns, the most commonly encountered being the little brown myotis, *M. lucifugus.* They come to life at twilight and swoop through the darkening night after the insects on which most of the bats in this hemisphere feed. Anyone can look for bats, since they occur, in a variety of species, east and west, north and south, though they do not venture north of the northern forests.

FAMILY DASYPODIDAE:
ARMADILLOS

FOR PROTECTION the skunk has adopted an evil smell, and the porcupine a coat of sharp spines, but the armadillo is covered with a coat of mail. The skin of animals is extremely versatile in its functions. It has produced the feathers of birds and the hair of mammals, in great variety of color and form. The horns of the pronghorn and the mountain sheep, and the furry velvet of deer antlers, are special adaptations of the hair-producing function of skin. It would be interesting to know by what long evolutionary steps the skin of the original armadillo ancestors developed, in lieu of the usual hairy coat, the hard shell that today covers most of the body of this strange burrowing mammal.

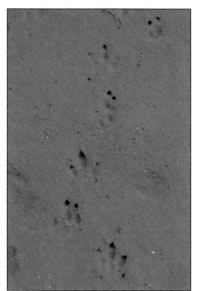

The nine-banded armadillo, *Dasypus novemcinctus,* is a product of South America but has come up from the tropics to inhabit Mexico, most of Texas, Oklahoma, and the entire southeastern United States. And still its range pushes north.

The walking trail of an exploring armadillo in southern Texas.

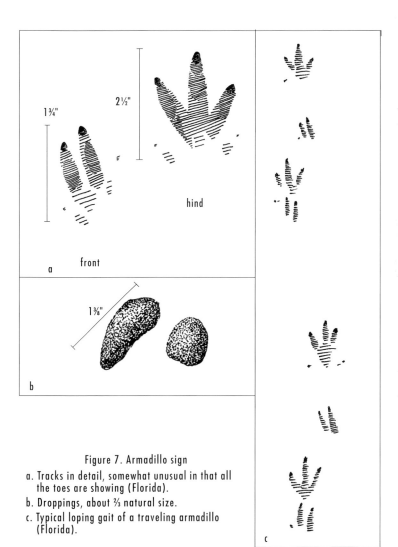

Figure 7. Armadillo sign

a. Tracks in detail, somewhat unusual in that all the toes are showing (Florida).
b. Droppings, about ⅔ natural size.
c. Typical loping gait of a traveling armadillo (Florida).

The tracks of the armadillo are most unusual among North American mammals. You are more likely to confuse the trail of an armadillo with that of a small alligator or some bird than with another mammal. In some surfaces the tracks have been referred to as "hooflike." On occasion, too, the reticulated imprint of the "shell" may be seen in the dirt. But most often only partial prints or nails register, making their identification that much more difficult. The front foot of the armadillo has four toes, but the two side toes are found on a higher plane than the two central, and therefore only register in deep substrates. The hind feet have five toes, but again the two outer toes are found on a higher plane. It is most common to find front tracks of only two toes and hind tracks of only three. Front tracks measure 1 ½–2 inches long and 1 ⅜–1 ⅝ inches wide. Hind tracks are 2–3 ¼ inches long and 1 ½–2 ⅜ inches wide.

The armadillo feeds on insects, but in doing so, like the bear, it eats a large quantity of dirt and other debris. Figure 7, b, shows the droppings, which are mostly composed of clay. They are generally round and marblelike. These clay droppings suggest those of elk and deer after those animals have eaten mud at a mineral lick. Armadillos form latrines which are most easily found in dry climates, where their scats hold form longest. At other times they bury their droppings.

Here is a most efficient digger that can excavate many burrows, 7 or 8 inches in diameter and from about 2 to 15 feet in length. It also uses natural cavities. Since it digs and roots for insects and excavates anthills for food, these disturbances are also evidence of its presence. Note that certain skunks also root after insects, and peccaries seek food underground by rooting. Accordingly, one should try to find tracks or droppings as aids to identification. Bears also disturb anthills, but ordinarily are not in the same habitat with armadillos.

FAMILY OCHOTONIDAE: PIKAS;
FAMILY LEPORIDAE:
HARES AND RABBITS

Figure 8. Rabbits and pika

a. Arctic hare.

b. White-tailed jackrabbit.

c. Black-tailed jackrabbit.

e. Pygmy rabbit.

f. Cottontail sp.

g. Pika sp.

d. Snowshoe hare. The Washington hare, at left, does not turn white in winter
as do other snowshoe hares.

EVERYONE IS familiar with rabbits, even if through nothing more than the Easter rabbit and Bugs Bunny. But the rabbit of the wilds is something more than the generalized Peter Rabbit of delightful stories. If you go forth as a naturalist, amateur or professional, you will find before you a story with great significance. A glance at figure 8 reveals a gallery of distinct rabbit personalities. You will find these linked to distinct territories in which, through the many centuries, each type of rabbit has "sunk to its roots," so to speak, and become an intimate part of the particular environment. I hope the following accounts will help to reveal this story of adaptation. In this group is included the diminutive mountaineer, the pika, or cony, a relative of the rabbit clan.

The rabbit track is distinctive as a track type, varying in size with the many kinds. The dropping also is very distinctive and varies in form less than in other animal species. Normally it is somewhat flattened and circular; we might call it a thick disk. Even the pika scats approach this form, except when the diet has been soft food. Figure 14 presents the rabbit type, as well as the differences in size characteristic of the several species. You will also note that there are some variations in each case.

Since the rabbit track is of a distinctive pattern, and the tracks of the different kinds vary chiefly in size, they are not presented at this point.

In the following individual accounts an attempt is made to show not only the characteristics of each group but also, collectively, variations that are more or less common to all.

COLLARED AND AMERICAN PIKAS

Pikas, of the genus *Ochotona*, are 6–8 inches long and found in the Rocky Mountain regions of the western states, southern British Columbia and Alberta, the Cascades and Sierra Nevada, and the mountains of interior Alaska and the Yukon. The northern species is called the collared pika, *O. collaris*, and the southern the American pika, *O. princeps*. This distribution reveals the way of life of this diminutive relative of the rabbits, which has chosen to spend its time among the jumbled blocks of a rockslide or in some other rocky habitat.

The pika, or cony, is hairy-footed like the rabbits, but it has short round ears and shows no tail. Pika tracks are most often seen in late, lingering snowdrifts, early fall snow, or mud at the edge of the water near their rock refuge. Figure 9 shows track patterns in somewhat fragmentary form. Note that the front feet have five toes, though often all do not appear in the track. The hind feet have only four toes, like the rabbit. Front tracks measure $^{11}/_{16}$–$^{7}/_{8}$

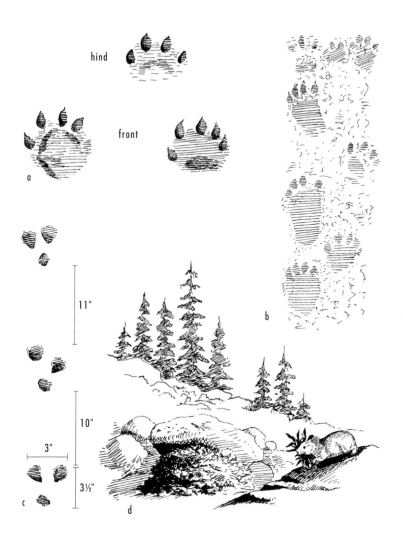

hind

front

a

b

11"

10"

3"

3½"

c

d

Figure 9. Pika sign, in Wyoming

a. Tracks in mud at margin of stream.
b. Fragmentary trail, shuffling gait, in rough snow.
c. Bounding gait.
d. Pika putting up hay for winter.

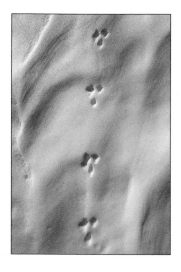

(Left) *The bounding trail of an American pika atop a mountain in Colorado.* (Above) *A pika's haystack in Colorado Rockies. It may be easier to locate a haystack than clear tracks in their preferred rocky terrain.*

inch long and $^{11}/_{16}$–$^7/_8$ inch wide. Hind tracks are $^3/_8$–1 $^1/_8$ inches long and $^3/_4$–1 inch wide.

One obvious sign of the pika is the distinctive call. If you are following a trail through rough rock and from somewhere nearby comes a small voice, a sharp little bleat crying *Enk!*, look around carefully and you may spy a small furry lump sitting still, or a little gray form running smoothly over the rock surface to disappear in a crevice. It will be the pika.

In the winter of 1921 I was driving a dog team up the Toklat River in the Alaska Range. I had not come to the higher mountain slopes, but was passing a tall bluff of the riverbank when I heard a familiar sound. I listened and heard it again, and recognized the pika's voice. It came from within that mass of snow. From its retreat inside, the animal had heard us passing out there in the Alaskan winter and, true to its nature, had given voice to curiosity or excitement.

The incident illustrates another important part of pika economy. It can spend the entire winter under the snow, for it has put by stores for that purpose. In the late summer you will find a pile of dried plants and twigs under an overhanging rock or beside a large boulder, often with fresh green food on top of the pile. There may be little bundles of plants drying nearby, not yet added to the winter store. And if you look around, you may see the old matted vegetation of former haystacks.

The droppings lie among the rocks and look like round pellets of black tapioca. If the animals have had succulent feed the drop-pings will be elongated. Compare figure 14, h and i.

PYGMY RABBIT

This pygmy of the rabbit clan, *Brachylagus idahoensis,* is a west-erner of restricted range. It has chosen the dry sagelands in an area including parts of Idaho, Oregon, northern California, and Nevada. And within this area it seeks out the tall sage growth. In northwestern Nevada, where so much of the sage is the low sparse kind, I would come to one of the thick patches of tall sage, *Artemisia tridentata,* looming up like a miniature forest on the flatland. There I would find the pygmy, which measures 8½–11 inches in length. I was able to approach close, and when the little gray rabbit fled it scuttled away among the heavy bushes and soon disappeared. It regularly seeks shelter in underground dens that it

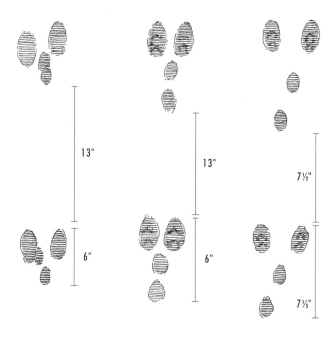

Figure 10.
Several track patterns of the pygmy rabbit,
at slow speed (northwestern Nevada, January 1942)

digs itself, the only rabbit in North America to excavate such retreats. For track patterns, see figure 10.

The droppings of this rabbit are so small that they approach a spherical form, rather than the disk (see figure 14).

COTTONTAILS

The eastern cottontail, *Sylvilagus floridanus,* is probably the best known of the rabbit family. But in one form or another, cottontails are native to nearly all of the United States and Mexico and extend into Central America.

The familiar tracks of this group of rabbits may be found equally in sagebrush deserts and the eastern woodlands, though of course in different species. You will even find the tracks in the outskirts of cities. In fact the tracks shown in figure 11 were drawn in Washington, D.C. You will note that they follow the usual rabbit pattern, in a special size and edition. The front tracks of eastern cottontails measure 1–1⅞ inches long and ¾–1⅜ inches wide. Hind tracks are 1¼–3¼ inches long and ⅞–1 13⁄16 inches wide.

Cottontails will gnaw twigs in winter and occasionally will eat the bark on orchard trees. It is not always easy to distinguish their tooth marks from those of rodents, but the marks are larger than those of mice. The twig shown in figure 11, d, is from Wisconsin.

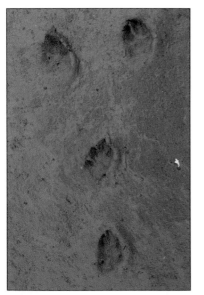

The end of the twig shows the effect of repeated bites, different from the clean cut on the small twigs eaten by the black-tailed jackrabbit (figure 15, d), or plants eaten by the snowshoe hare.

Eastern cottontail tracks in the typical bounding pattern, coastal Virginia.

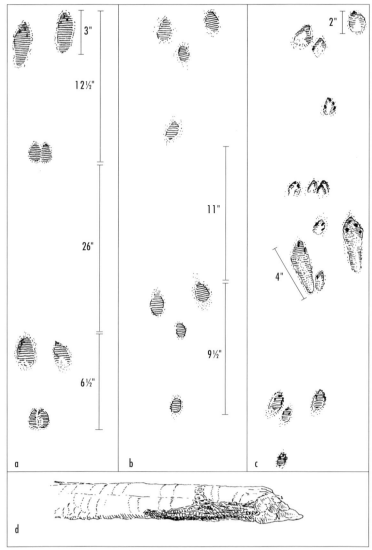

Figure 11. Eastern Cottontail

a. A medium-paced bounding gait.
b. A slow bounding gait.
c. Cottontail stopping, then bounding on. The bounds were from 4½–6 in.
d. Twig gnawed by cottontail, about natural size.

Cottontails will often rest in the usual "form," under a sage bush or similar shelter in the West, or some other bush or tangle of vines or a brush heap in the East. They will hide in hollow logs, stumps, burrows in the ground, or a crevice in a rock.

These rabbits are only occasionally found in the same places as the snowshoe hare, so their tracks and droppings are less likely to be confused. Reference to figure 14 will show that cottontail droppings are much smaller than those of jackrabbits, whose territory they often share. However, determining a rabbit species only from a small number of pellets may be misleading.

MARSH AND SWAMP RABBITS

The marsh rabbit, *Sylvilagus palustris,* inhabits marsh country from Virginia to Florida. The somewhat larger swamp rabbit, S. *aquaticus,* occupies wooded swamps, marshes, and river bottoms from western Kentucky and Georgia westward into Texas. Related species occur in Mexico and Central and South America.

Although these specialized rabbits are at home in water and choose wetlands to live in, and sometimes readily plunge into water to escape pursuit, they really have many habits in common with the cottontail. For example, they rest in forms in brier patches, thickets, and similar shelter, and enter burrows of the gopher tortoise—hollow logs and stumps and hollow trees. When danger approaches they will climb up some distance inside these trees, as I have found cottontails doing in Minnesota. In short, all of these rabbits seek some kind of cover for concealment, using whatever is available in their chosen habitat. The marsh rabbit finds shelter in tall grass and cattails. It will also venture onto floating vegetation, as one would expect from such a water-loving animal. Marsh rabbits have definite runways in the vegetation,

The unique walking trail of a marsh rabbit in southern Florida.

and also on old mossy logs, I am told. The droppings are similar to those of the cottontail. I doubt if they could be distinguished consistently. However, if you find them on logs, along runways, or on floating material in water of the swamp areas, you may conclude the marsh rabbit or swamp rabbit has been there. In fact it is quite characteristic of the swamp rabbit to defecate on elevated surfaces, such as logs; I've yet to see a cottontail do this. Also consider the territory, as the cottontail normally seeks higher ground.

Figure 12. Tracks and scats of marsh rabbit
a. Close-up of tracks of walking marsh rabbit.
b. Walking track pattern of marsh rabbit.
c. Droppings of marsh rabbit, natural size (Everglades National Park).

One investigator, Ivan R. Tomkins, discovered that *walking,* with alternate steps, is one gait of the marsh rabbit — a significant departure from the almost universal hopping of the rabbits. Figure 12, a, is adapted from photographs kindly furnished by Mr. Tomkins; the distance between steps measures 3½–6½ inches.

SNOWSHOE HARE

Lepus americanus occupies a tremendous area — the entire transcontinental coniferous forest, including Canada, Alaska, the northern states, and the Rocky Mountains and Pacific Coast forests as well. This hare has found the temperate climate best, but has gone as far north as it can find forest. It is well named, for its hind toes spread to form a broad "snowshoe" surface on the snow (see illustration). In figure 13 note the wide shape of the tracks that is a result of this feature. Around the cabin in Wyoming where I am writing these lines, the patterns a and b are everywhere on the surface of the snow (which is 4 feet deep). These animals are foraging for twigs — and the vegetable peelings we throw out for them. One that had been frightened by a coyote went off in great bounds, as in c. Front tracks measure 1⅞–3 inches long and 1⅛–2¼ inches wide. Hind tracks are 3¼–6 inches long and 1⅝–5 inches wide.

In the north woods, when these rabbits become numerous, they form well-established trails in the snow. During a big rabbit year in the Hudson Bay country, 1915, the Indians snared great numbers of them in these trails. Since several goshawks and barred owls were snared in the same region, this indicated that they too recognized good rabbit-hunting grounds, and apparently at times actually used the trail for the chase.

Tight tracks of a snowshoe hare in a typical bounding gait in central Idaho.

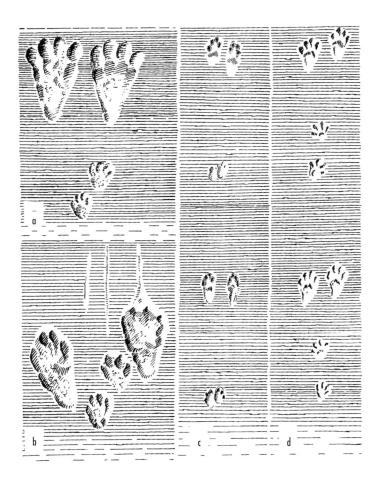

Figure 13. Tracks of snowshoe hare in snow

a. Snowshoe hare, *Lepus americanus,* slow bound, in snow. Length of track pattern about 11 in.; hops, about 14 in. (Olympic Mountains, Washington).

b. Snowshoe hare of Rocky Mountains, slow hopping. Length of track pattern about 10 in.; hops, about 10 in. (Wyoming).

c. Rocky Mountain snowshoe hare speeding. Track patterns about 24 in.; leaps, 38–67 in. (Wyoming).

d. A hare speeding, and showing characteristic track splay found in deep snow conditions. Track patterns 20–22 in.; leap, 66 in. (Olympic Mountains).

Sometimes you will find twig cuttings left by the snowshoe. A lodgepole pine beside my cabin in Wyoming was blown over in a high wind, and during the winter these rabbits trimmed all the twigs within reach. Or you will find areas where the hares have fed upon the bark of trees. Characteristic of all members of this family, when feeding on the cambium of woody plants, hares gouge far beyond the bark into the woody center.

Once on a gravel bar of the Snake River in Wyoming I found some lupine plants that had been nibbled. What rodent had cut these? The slant cuts on the eaten stems meant rodent work, of course. I thought of setting a live trap to find out. But I went back for another look and this time noticed the telltale round scats of the snowshoe hare. There were several beside each plant that had been cropped.

Usually the snowshoe will rest at the base of a tree or bush, but often takes shelter under a log or brushy place. In winter, when the snow is deep, some excellent tunnels form under logs and limb tangles, sheltered above by a great load of snow, and the hare finds refuge there, far out of sight. Like the cottontail it will also find refuge under a log cabin.

ARCTIC AND ALASKAN HARES

The Arctic hare has become specialized for living in the high North, from Greenland to the Northwest Territories. Chiefly it inhabits tundra, or open country, but on the shores of the southern Bering Sea you are likely to find *Lepus arcticus* in the alder brush. One variety lives as far south as Newfoundland. Like our familiar white-tailed jackrabbit, which it resembles in many ways, the Arctic hare turns to white in winter. In northern Greenland and on Ellesmere Island and adjacent parts of the Arctic Archipelago, the hares don't even bother to turn gray for the short summer they have, but remain white the year round. The Alaskan hare, *L.*

Figure 14 (opposite)
a. Arctic hare (Bristol Bay, Alaska).
b. White-tailed jackrabbit (Jackson Hole, Wyoming).
c. European hare (New Zealand).
d. Black-tailed jackrabbit (Nevada).
e. Snowshoe hare (upper, Minnesota; lower, Wyoming).
f. Cottontail (upper, Nevada; lower, North Dakota).
g. Pygmy rabbit (Nevada).
h. Pika. Soft type from succulent feed (Alaska).
i. Pika. Hard type, from dry feed of fall and winter (Alaska).

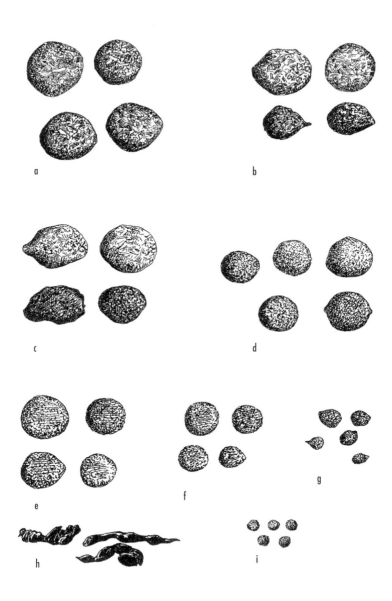

Figure 14. Rabbit and pika scats, natural size

othus, is the western counterpart to the Arctic hare and weighs 9 or 10 pounds. Some even heavier weights have been reported, so this animal is in many respects quite comparable to our white-tailed jackrabbit of the open plains. In general form the tracks are similar to those of the white-tailed jackrabbit shown in figure 17. Front tracks of Arctic hares measure 2¾–3¾ inches long and 1¾–2½ inches wide. Hind tracks are 4¼–8 inches long and 2⅜–3⅞ inches wide.

The Arctic hare has a remarkable habit that produces tracks very different from the usual rabbit trail. Occasionally it will hop along on the hind feet only, kangaroo style, for some distance before dropping back to all fours. Travelers who have witnessed this say it is a striking spectacle to see a sizable group of these animals take off all together on their hind toes. Colonel John K. Howard, who photographed these animals in Greenland many years ago, kindly made available some movie film, on the basis of which the sketch on page 54 was made.

BLACK-TAILED AND ANTELOPE JACKRABBITS

The black-tailed jackrabbit, *Lepus californicus,* is primarily the rabbit of the western sage and cactus country, though it has occupied grassy plains as well—from Nebraska, Kansas, and Oklahoma west to the Pacific and from Washington south into Mexico. The related antelope jackrabbit, *L. alleni,* ranges up from Mexico into New Mexico and Arizona. Neither of these turns white in winter. The front tracks of black-tailed jackrabbits measure 1⅞–2½ inches long and 1¼–1¾ inches wide. Hind tracks are 2½–5⅝ inches long and 1⅜–2½ inches wide.

The splayed hind tracks of a black-tailed jackrabbit in the grasslands of Colorado.

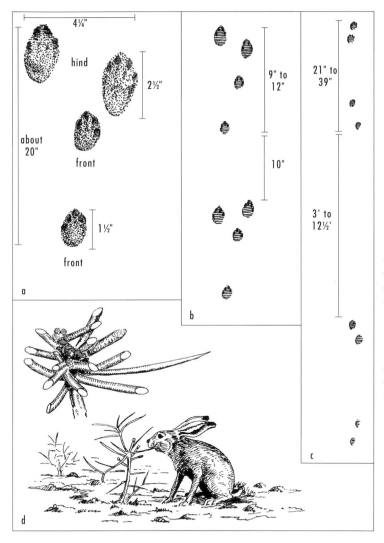

Figure 15. Black-tailed jackrabbit, from Nevada

a and b. Typical tracks in sand, at slow speed. Width, or straddle, 4¼ in.

c. Speeding, in snow.

d. Feeding on "crown of thorns" *(Koeberlinia spinosa)* in the Texas desert. Cut twigs are shown above.

In parts of the western range both black-tailed and white-tailed occupy the same areas, hence their tracks could be confused. However, in general the tracks of the black-tailed jackrabbit are considerably smaller than those of the white-tailed.

You will notice that the rabbit track patterns are similar in all the species, the hind feet coming ahead of the front feet in the normal gaits. The reason is that the rabbits are primarily bounders, even when moving slowly. One day my son and my brother and I, traveling through Arizona, came on a black-tailed jackrabbit at close range. My brother took a series of motion pictures. To our surprise the rabbit began *walking* away, quite unrabbitlike. Had he become stiff lying in his form? Later he hopped away vigorously enough, but he did walk; the illustration on page 54 is a sketch taken from the motion picture. It is interesting to refer to figure 12, in the marsh rabbit section, and the drawing of Arctic hares hopping, also on page 54.

I have not seen the tail mark in the snow described by Seton, though sometimes there are drag marks of the toes, similar to those of the snowshoe hare in figure 13, b.

Note the slight differences in size of droppings, in descending order, of the Arctic hare and white-tailed and black-tailed jackrabbits (figure 14).

As with some of the other rabbits, this species occupies forms rather than burrows. You will find these slightly hollowed resting places under bushes, in a grassy place, by a rock, or any handy spot that gives a little protection.

The black-tailed jack feeds on a variety of desert shrubs and cacti. You will find prickly pear cactus gnawed by these as well as other animals. Note the slant cut characteristic of rodent and rabbit work in the upper part of d, figure 15. A deer would leave a "pinched-off" effect.

WHITE-TAILED JACKRABBIT

This big rabbit of the plains, *Lepus townsendii*, is found from the prairies of the midwestern states and southern Canada, west through the sagelands to the high mountain slopes of the Rockies, Cascades, and Sierras. It is almost as big as the Arctic hare, weighing 7 or 8 pounds and more. Here is an animal that looks much like the Arctic hare, but it has found the northern plains and high western slopes more to its liking than the wastes of polar regions.

The white-tailed jackrabbit is a good subject with which to illustrate some of the difficulties in tracking. Figure 16 shows well-known tracks and some track patterns in snow. In my experience the leaps vary from 1 or 2 feet in slow hopping up to 9 feet

The bounding, then pausing, then bounding trail of a white-tailed jackrabbit in the Colorado Rockies.

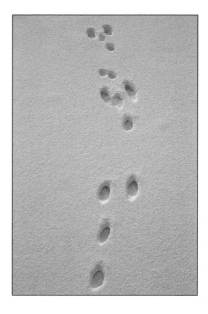

or more at high speed. Others have reported as much as 12 and 20 feet at extreme speeds.

You do not always find perfect tracks. As boys in Minnesota, my brother and I used to hunt jackrabbits by tracking. Sometimes in bad weather we would lose the trail when the rabbit had crossed an area of hard crusted snow where the wind was sweeping away all loose material. There would be no indications of the long hind foot, but the typical pattern of a few claw marks on the crust still spelled rabbit and we could recognize the speeding pattern of figure 16, e, or the slowing pattern of a. Front tracks measure 2⅛–3¾ inches long and 1½–2⅝ inches wide. Hind tracks are 2½–6¾ inches long and 1⅝–3⅜ inches wide.

Figure 16, c, by itself would suggest coyote. But remember that substrate conditions play tricks with footprints. Examine the pattern closely and its size will tell you the true story. It is important not to expect perfect tracks or to rely on one set of known characters. Even in good mud tracks of the beaver, for example, you rarely find all five toe prints of the front foot, although you know that is the number it really has. And so you learn to add previous knowledge and judgment to what you find recorded in mud or snow. For instance, any rabbit track in Greenland or the polar region *beyond the limit of trees* would mean Arctic hare and not jackrabbit. If you find a certain type of dropping in the trail of a northern forest where it is known that there are no bobcats, you may conclude that a lynx has passed that way, but if bobcats also live in the forest, you are not so sure. A store of dried grass and leaves heaped up at the base of a willow in the tundra area of the Alaska Range means the singing vole, the little "hay mouse" that puts up winter supplies. A similar pile of hay up in the rocks more likely means pika, or cony, and certainly would in the Rocky Mountains.

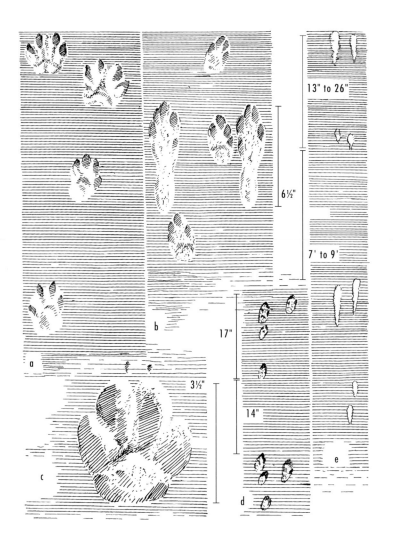

Figure 16. White-tailed jackrabbit tracks in snow (northwestern Wyoming)

Figure 17. The zigzags in the trail spoke of an attack, and a snowy owl nearby suggested the foe (Moorhead, Minnesota, January 20, 1918).

Figure 16 (opposite)

a. Typical bounding track pattern, only the toes of hind feet touching. Length of pattern 19½ in.; length of hind print (in front) about 3½ in.; length of front prints (in rear of pattern) 3 in.
b. Sitting track pattern, hind heels down, the front tracks of last leap showing between the hind tracks. Greatest width of this pattern, or straddle, is 8 in.
c. Hind-foot tracks without heel mark, showing general resemblance to coyote track.
d. Pattern in moderate speed.
e. Speeding.

It is this sorting of facts and previous experience, combined with the imperfect evidence in the snow or mud before you, that makes success in reading nature's prints, and makes of such reading an exhilarating experience.

One winter day in Minnesota I was following a jackrabbit track across the snowy fields. Presently I came on some peculiar maneuvers. The trail suddenly broke in several *zigzag* turns. There were no other tracks there on the snow. Had the rabbit been seized by a giddy playful mood, or a nightmare? The trail straightened out as I followed along, puzzled, and very soon I came to a haystack. There sat a snowy owl, one of those winter visitors from the far north. The story then became clear. The owl had made several swoops at the running rabbit, the rabbit had nimbly dodged each attack, and for some reason the owl gave up and went over to perch on the haystack (see figure 17). A flying bird leaves no track, but it can create a mystery in the snow!

The white-tailed jackrabbit digs no burrow, but rests in the traditional "form"—a hollow in the snow or in the grass, at the base of a bush or tree, beside a rock, or the entrance to a badger hole (see above). In cultivated areas this rabbit often snuggles down beside a clod in a plowed field. If the snow is deep enough, it may dig a shallow hole in the snow where it can crawl in out of the weather.

Figure 18. White-tailed jackrabbit speeding away from its form in the snow

This European immigrant, *Lepus europaeus,* about the size of the white-tailed jackrabbit, has been introduced in some of the northeastern states, from New Jersey and eastern Pennsylvania and New York to Connecticut and Ontario. My own experience with this hare was not in America, but in the Southern Alps of New Zealand. There, on the slopes of Mount Sebastopol after a recent snow, I found the tracks of this animal—an immigrant from Eu-

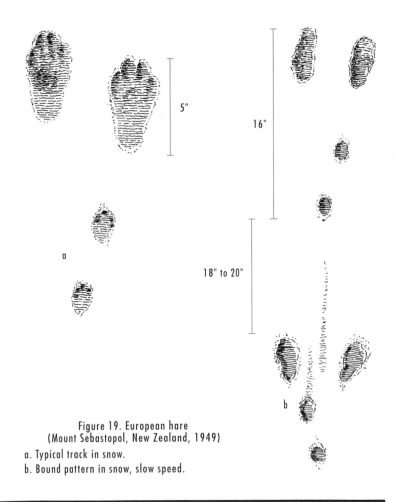

Figure 19. European hare
(Mount Sebastopol, New Zealand, 1949)

a. Typical track in snow.
b. Bound pattern in snow, slow speed.

rope here, too. It was living on a high open mountain slope, where it had been trimming the lower portions of native shrubs, while the introduced chamois and tahr had been cropping the higher twigs. The hare cuttings were diagonal, like those shown in figure 15, d.

In figure 19 are shown the tracks, only slightly smaller than those of the white-tailed jackrabbit. The droppings (see figure 14, c) seem fairly comparable in size to those of the whitetail.

ORDER RODENTIA: RODENTS

THE ANIMALS WE classify as Rodentia, the chisel-toothed animals that gnaw and live on vegetation, are extremely numerous and extremely varied. They include large creatures such as the beaver, agouti, paca, and marmot, as well as the hordes of mice of many species. There are those that eat leafy vegetation, those that live mainly on seeds, even those that fell trees and eat bark. These varied habits are reflected in the signs they leave in the mud or snow, or in the flora.

In figures 23, 24, 25, and 26, representative tracks and droppings of the rodents are assembled for handy reference. So far as possible they are drawn to scale on each page. The pages on scats are presented about natural size.

For more detailed study refer to the discussions of the species which follow.

APLODONTIA OR MOUNTAIN BEAVER

In the Pacific Coast mountains, particularly in the rain forest environment, from the redwoods of northwestern California to southern British Columbia, lives an interesting rodent, *Aplodontia rufa,* the aplodontia, often called the mountain beaver, or, occasionally, the sewellel. It is not closely related to the beaver but is a chunky, virtually tailless animal nearly as large as a muskrat. In terms of evolution, it is the oldest rodent in North America.

The tracks of the aplodontia share many characteristics of rodents. They have four toes on the front feet and five on each hind. The claws of the front feet are well developed for digging, and very long. Also their hind tracks appear like tiny hand prints, yet their innermost toes are strangely thinner and pointier than the others. Front tracks measure ¹⁵⁄₁₆–1 ⁹⁄₁₆ inches long and 1–1 ⁹⁄₁₆ inches wide. Hind tracks are ⅞–1 ⅝ inches long and ¹⁵⁄₁₆–1 ½ inches wide.

The walking trail of an aplodontia in a mud puddle in the forests of coastal Washington.

As you travel a coastal forest trail, or particularly in logged-off areas, you may come upon little piles of vegetation—ferns and other plants—apparently laid out to dry. They may be on the ground near burrows, but are often on logs. These represent the haymaking of the aplodontia, generally thought to be for nest material as well as for food storage, and suggest the hay storage of the pika. But remember that the pika piles will be in rock slides, or similar rocky habitat, and not in the rockless forests.

Furthermore, near these plant piles you will find the aplodontia burrows, in diameter about 4–8 inches. They form an irregular network, with several openings. There may also be a "dump pile" of earth thrown out from burrows. In the Olympic Mountains I have come across these burrows along the trails, sometimes caved in, apparently wet and soggy in the rain-soaked woods. The animals have their well-constructed nest within, so they do not mind water.

Nearby you may find limbs nipped off bushes or small saplings. It is reported that they may even cut down a small sapling. They will climb among the limbs to nip off higher ones.

In winter, like some other rodents, the aplodontia will move about under deep snow. In spring the earth cores are revealed,

Figure 20 (opposite)
a. Above, the hind left track; below, left front track (note long claws) (Washington).
b. Scat, about natural size. Structure of droppings depends upon size of animal and diet; color dark green to black if fresh, turning gray-green to tan with age (drawn by Carroll B. Colby from specimens in the American Museum of Natural History, New York).
c. Walking trail in mud. Strides vary from 4 to 7 in. (Washington).

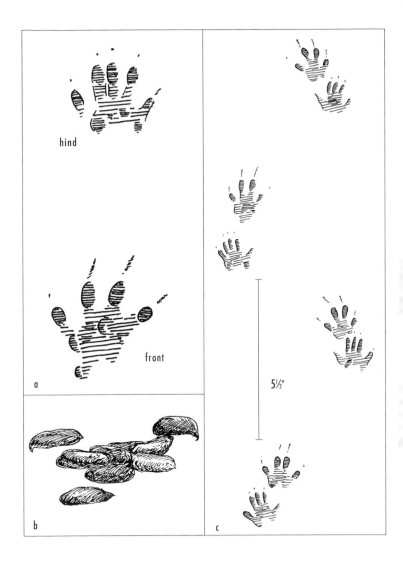

hind

front

a

b

c

5½"

Figure 20. Aplodontia *(Aplodontia rufa)* tracks and droppings

consisting of the excavated dirt pushed into snow tunnels after the manner of the pocket gopher. These cores are much larger, however, being 6 inches or more in diameter, corresponding with the burrow size.

I have never heard the aplodontia utter a sound, and it has always seemed to be a silent dweller of the wet burrows of the dark forest. A few observers have described it as producing a series of whistles, a fact that seems to be corroborated by the testimony of early Indians. Since it is active at dusk and during the night, that is the time one might expect to see and hear this animal.

CHIPMUNKS

Chipmunks are familiar to most people. The large eastern chipmunk, *Tamias striatus,* is found in the eastern half of the United States (except the southeastern corner), from Louisiana, Iowa, and Minnesota east and north into southeastern Canada and to the Atlantic Coast. To the west of this range are all the diverse species, large and small, dark and pale, that may belong to a separate genus, *Neotamias* (the issue is still being debated), occupying mountains, forests, and deserts. But wherever found, of whatever species, the chipmunk acts the same and is the same pert character with those who take note of such things. Following the *Peterson Field Guide to Mammals,* all chipmunks are here referred to as *Tamias* species.

These sprightly members of the squirrel family are mostly terrestrial, though they will readily climb trees. Their tracks may be found in mud, dust, sand, or snow. Snow tracks, however, will be seen only in the fall, early spring, or during warm spells midwinter, for in snow country these animals have stored up food and remain for many months underground.

Figure 21, a, shows the tracks of a western chipmunk in mud, in which the animal ran mostly on the toes; b shows tracks in dust in the same locality, in which the hind heels were put down. These track patterns vary in width from 1 ½ to 2 ⅞ inches, depending on the spread of the hind toes. The length of the track pattern, from rear of hind tracks to toes of front tracks, varies from

Figure 21 (opposite)
a. Tracks in firm, wet mud (Wyoming).
b. Tracks in dust, heel of hind foot showing (Wyoming).
c. Scats of three species of chipmunk (top down: *Tamias alpinus,* Wyoming; *T. striatus,* New York; *T. dorsalis,* Arizona).
d. Bounding trail in wet snow (Wyoming).
e. Bounding trail in new snow (Wyoming).

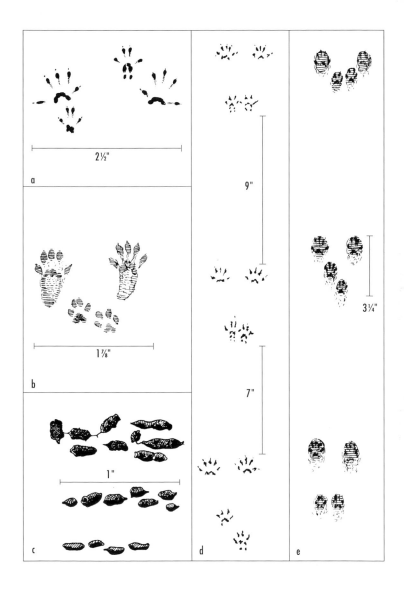

Figure 21. Chipmunk tracks and scats

Bounding tracks of an eastern chipmunk along a muddy riverbank in Connecticut.

about 1¾ to 5 inches. The leaps may be from 5 to about 23 inches. Figure 21, d, most nearly expresses the pert, neat, and agile movements of the chipmunk.

Again it should be noted that in the leaping gait the hind feet fall in front of the forefeet tracks, and in the case of the chipmunk the two pairs of tracks, hind and front pairs, are parallel with each other in most instances, though sometimes the two forefeet tracks are one behind the other, as with the ground squirrel.

The front tracks of the large eastern chipmunk measure ¾–1 inch long and ⁷⁄₁₆–⅞ inch wide. The hind tracks are ½–⅞ inch long and ⅝–¹⁵⁄₁₆ inch wide. The front tracks of the smaller least chipmunk, found in the West, measure ⅜–¹¹⁄₁₆ inch long and ⁵⁄₁₆–⁹⁄₁₆ inch wide. The hind tracks are ½–¹¹⁄₁₆ inch long and ½–¹¹⁄₁₆ inch wide.

The droppings are to be found in figure 21, c. These particular samples show considerable differences in size and form. Much of this difference can very well be due to the type of food eaten and its quantity. They are not correlated with the size of the three species of chipmunks here represented.

All the chipmunk burrows that I have seen have had simple, unobtrusive openings in the ground, without an evident earth dump at the entrance. In fact, the cleanliness of these entrances is one of their identifying characteristics. The diameter is approximately 2 inches.

The chipmunk's diet is extremely varied, but mostly consists of nuts, berries, and seeds of various kinds. The eastern chipmunk feeds on a diversity of nuts and grains, such fruits as wild cherries and raspberries, mushrooms, insects, some carrion, and occasionally invertebrates, reptiles, amphibians, and even other small mammals. In Minnesota we used to find quantities of dry shells

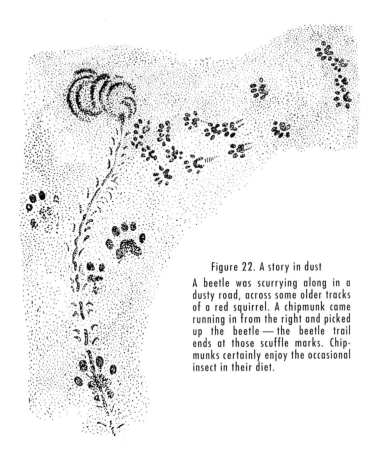

Figure 22. A story in dust

A beetle was scurrying along in a dusty road, across some older tracks of a red squirrel. A chipmunk came running in from the right and picked up the beetle — the beetle trail ends at those scuffle marks. Chipmunks certainly enjoy the occasional insect in their diet.

of the basswood nuts, opened on one side. We saw chipmunks pack several white oak acorns into their cheek pouches, and marveled at their capacity. Chipmunks took advantage of hollow trees and logs, and brush heaps.

In the West I have found the chipmunks feeding on grass seeds, and especially the dandelion heads, just when the calyx has closed after blooming. On a log or stump or rock, or on the ground, you may find remains of such feasts — a little pile of dismembered grass-seed heads, several opened dandelion heads, or remains of other favorite plants. These are sure chipmunk signs, as the defensive strategy of chipmunks is to feed up high where sight or sound of potential predators is most easily detected.

The chipmunk has a call, but I have not found great variety in its repertoire. Usually the call has been represented as *chock, chock, chock* repeated to various lengths. It may be only two or three *chocks,* or these may be repeated for a considerable length of time. Their other common calls are loud whistles given in alarm as they retreat below cover while danger approaches.

YELLOW-BELLIED MARMOT, HOARY MARMOT, AND WOODCHUCK

The yellow-bellied marmot, or rockchuck, *Marmota flaviventris*, is the common marmot of the western United States, living mostly in and near the Rocky Mountains, the Cascades, and the Sierras. The hoary marmot, *M. caligata* (with three other species, *M. olympus, M. vancouverensis,* and *M. broweri*), is the large marmot of the high mountain country, most often found in the mountains of western Canada and Alaska but appearing also in some of the mountains of western Montana, Idaho, and Washington. The common woodchuck, *M. monax,* is the lowland marmot and is found in the eastern half of the United States (except the south-ernmost states), across the southern half of Canada (except the western coast), and on into easternmost Alaska.

The tracks of all these marmots are quite similar, though slightly larger in the case of the big hoary marmot. For example, note in figure 27 the similarity of the track patterns of the Nevada yellow-bellied marmot, *M. flaviventris* (e) and the Minnesota woodchuck, *M. monax* (f). Measurements of the individual tracks were about the same. Note also the similar measurements of the front tracks in a and d, yellow-bellied and woodchuck, respectively.

Front (above) and hind tracks of a yellow-bellied marmot in Yellowstone National Park, Wyoming.

Yellow-bellied marmot—*front*, 1¾–3 in. long, 1½–2⅜ in. wide; *hind*, 1¾–3¼ in. long, 1⅝–2½ in. wide

Woodchuck—*front*, 1¾–2¾ in. long, 1–2⅛ in. wide; *hind*, 1⅝–3⅛ in. long, 1⅜–2 in. wide

In these walking gaits the hind foot tends to register in the front track more or less, as with so many other mammals. In figure 27, g, and figure 28, c, are shown the variations in the running gait of the Wyoming yellow-bellied marmot, or rockchuck, which would be similar to those of the eastern woodchuck.

Note that the front foot, rodentlike, has four toes, big enough to show in the track, and the hind foot five. Also, the heel of the hind foot very often does not touch the ground, so that in some cases the front track appears longer than the hind track. However, in full plantigrade travel the heel of the hind foot bears down, as shown in some of the figures.

Tracks of the Wyoming yellow-bellied marmot on firm mud, as shown in figure 27, c, illustrate how the front and hind prints sometimes intermingle. The toes of the respective tracks can nevertheless be discerned. The width of the trail, or straddle, varies from an extreme of 6½ inches down to about 3½ inches.

When alarmed, marmots move in bounding leaps and leave similar patterns to the ground squirrels. The trail width of such bounds varies from 4¾ to 8 inches across.

The great variety in marmot droppings, consisting of vegetation, is indicated in figure 28, b. The inch scale in the figure will indicate roughly the sizes. The bottom sample illustrates the occasional soft type resulting from lush succulent food. The elongated one, too, failed to form the shorter, more normal segments.

Where do we look for these marmots? Let us consider the familiar eastern woodchuck

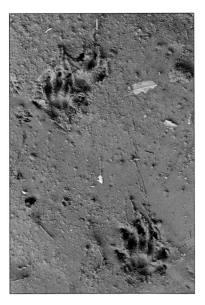

The walking trail of a woodchuck and the much smaller tracks of an eastern chipmunk along a river in Massachusetts.

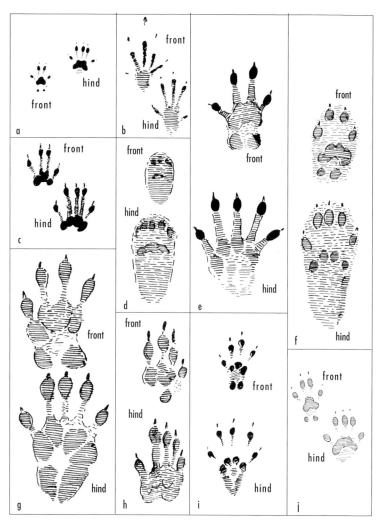

Figure 23. Tracks of marmots, squirrels, and gopher

a. Least chipmunk.
b. Northern pocket gopher.
c. Douglas's squirrel.
d. Northern flying squirrel.
e. Eastern fox squirrel.

f. Eastern gray squirrel.
g. Marmot.
h. Prairie dog.
i. Uinta ground squirrel.
j. Abert's or tassel-eared squirrel.

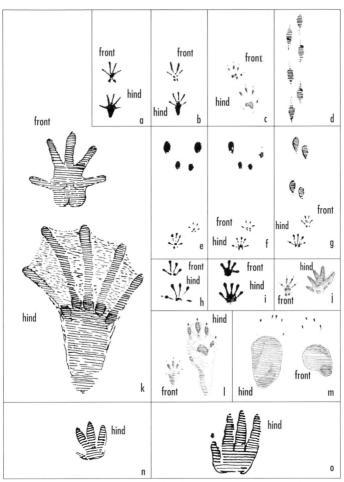

Figure 24. Some rodent tracks, drawn approximately to scale

a. Hispid cotton rat.
b. Rice rat.
c. Pocket mouse.
d. Lemming.
e. White-footed mouse.
f. House mouse.
g. Meadow vole.
h. Norway rat.
i. Woodrat.
j. Common muskrat.
k. American beaver.
l. Kangaroo rat.
m. North American porcupine.
n. Central American agouti.
o. Paca.

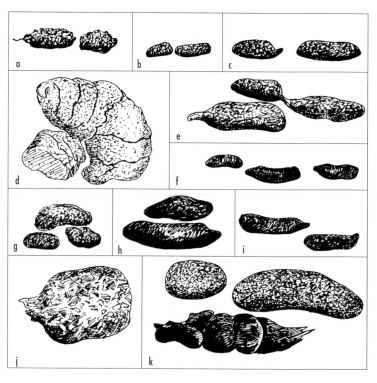

Figure 25. Rodent scats

a. Gray squirrel.
b. Pocket gopher.
c. Ground squirrel.
d. Marmot.
e. Prairie dog.
f. Woodrat.

g. Muskrat.
h. Aplodontia.
i. Norway rat.
j. Beaver.
k. Porcupine.

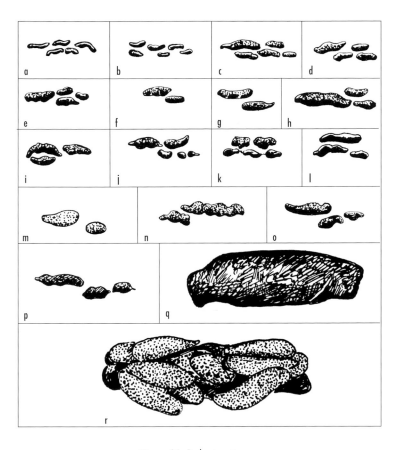

Figure 26. Rodent scats

a. Red-backed vole.
b. Pocket mouse.
c. Meadow vole
d. Kangaroo rat.
e. Collared lemming.
f. Brown lemming.
g. Grasshopper mouse.
h. Cotton rat.
i. Rice rat.
j. Jumping mouse.
k. White-footed mouse.
l. House mouse.
m. Abert's (tassel-eared) squirrel.
n. Red squirrel.
o. Flying squirrel.
p. Chipmunk.
q. Agouti.
r. Paca.

first. Mostly we find the woodchuck in the meadows, especially at the edges, but also in adjacent woods. In my experience, at least, they have been chiefly in open meadows and farm country, and are very often seen in the grassy edges along highways.

Woodchucks dig their own dens. Sometimes you will find the pile of excavated earth just outside the entrance, but other openings may be flush with the ground, with no sign of excavated earth. These flat, hard-to-find entrances are used as escape hatches when predators appear with little warning. A woodchuck peered at me from just such an obscure burrow entrance, on the bank of the Red River in Minnesota. In those same woods we often found woodchucks in a hollow tree, with the opening at ground level, or in hollow logs. Often the woodchuck would be resting on a convenient place on a slanting tree trunk, with a convenient cavity for refuge. Woodchucks do occasionally climb trees.

The woodchuck's voice is somewhat varied. When disturbed or threatened, it will chatter or whistle with a sort of trill and grit its teeth. But it has a more elaborate whistling accomplishment variously described by observers and some interpreters as its song.

Francis H. Allen wrote of this song in the *Bulletin* of the Massachusetts Audubon Society for November 1941:

> My notes made in Vermont, June 18, 1895, read: "Heard a woodchuck whistling just above the pine woods. I had never heard the note before and thought at first it was some rare bird unknown to me, but that was when he was some distance off. When heard near at hand the sound was absolutely startling. It begins very abruptly with a loud shrill short whistle followed immediately by another similar whistle not quite so loud and then by a succession of rapidly delivered softer and more liquid notes in a lower key. Quite a pleasing song on the whole. At a distance only the first two notes are heard." Though I am quite sure that I have heard the song several times since then, the only other record I find among my notes is "Sept. 18, 1904. Dover, Mass. Heard a woodchuck whistling in a mellow tone."

In March 1953 I visited the Trailside Museum at the Cook County Forest near Oak Park, Illinois, where there was a pet woodchuck. The young lady in charge called it up to the wires of the enclosure, stroked its face, and said, "Sing!" The woodchuck seized the wire in its teeth and issued a prolonged, piercing whistle in a monotone just as long as the young lady kept her fingers about its muzzle.

The so-called rockchuck, or yellow-bellied marmot, of the western states is well named, for it inhabits the high mountains, among the rockslides and cliffs, as well as some of the valleys, and

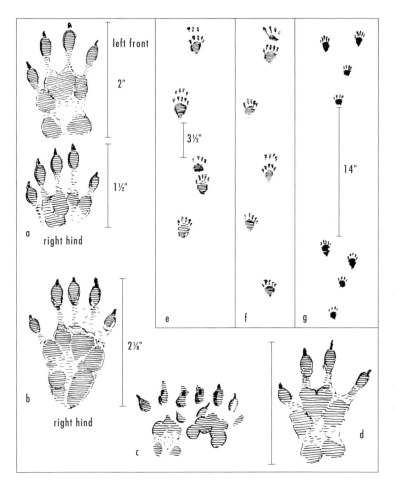

Figure 27. Tracks of yellow-bellied marmot and woodchuck

a. Front and hind tracks of yellow-bellied marmot, in sand, heel of hind foot not showing (Yellowstone National Park).

b. Right hind track, in mud, showing heel (Wyoming).

c. Intermingled front and hind tracks, on firm mud (Wyoming).

d. Left front track of the eastern woodchuck (Michigan).

e. Walking gait of yellow-bellied marmot (Nevada).

f. Walking/trotting gait of woodchuck (Minnesota).

g. Bounding gait of yellow-bellied marmot (Jackson Hole, Wyoming).

Figure 28. Yellow-bellied marmot (rockchuck)
a. In a favorite habitat, with den in rock cave (Wyoming).
b. Six samples of droppings, showing extremes in variation.
c. Variations in bounding and loping track patterns (Teton Mountains, Wyoming).

in lava beds. Its favorite retreat is among the rocks, though I have found its den also in the ground, much like that of the eastern woodchuck. I have never heard it "sing" like the woodchuck. The familiar call, undoubtedly an alarm call, is a short sharp whistle, a sound that carries a long distance.

The hoary marmot has also sought the high mountains, and its habits are much like those of the yellow-bellied. However, its alarm call is distinctive, being a more prolonged clear whistle, quite different from the brief, abrupt call of the "rockchuck." Another call has been reported, a series of rapid whistled notes. No doubt there will be found a more varied repertoire when the animal is more intimately studied.

GROUND SQUIRRELS

Ground squirrels, terrestrial members of the squirrel family, are in the genus *Spermophilus;* the closely related antelope ground squirrels are in the genus *Ammospermophilus.* Mostly western in distribution, they occupy open grasslands and meadows at all elevations from Michigan, Ohio, Indiana, Missouri, Oklahoma, Texas, and northern Mexico, west to the Pacific Coast, and northward through the western half of Canada and Alaska.

Ground squirrels vary greatly, and some of them are very striking. Some have stripes, some a finely speckled coat, lines of spots, or plain gray or brownish hair. There is great variety in color pattern and size in this group, but the track pattern is similar throughout. They live on vegetation but are also carnivorous to a considerable extent and will eat birds' eggs and birds if they can catch them. They feed readily on carrion,

The bounding track pattern of an Arctic ground squirrel in central Alaska.

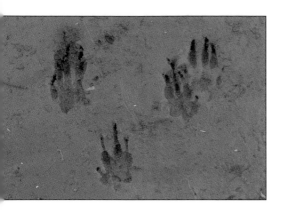

even the bodies of their own kind. Generally they have a whistle of some kind, which varies with the species. A few major types will be considered briefly.

The thirteen-lined ground squirrel, *S. tridecemlineatus,* of the central part of the continent, is the very slender one, adorned spectacularly with long rows of spots. In Minnesota I found that many of its burrows, about 2 inches in diameter and usually without an earth mound at the entrance, were so shallow I was able to excavate them with my hands to find the animal within. Front tracks of the thirteen-lined ground squirrel measure $^{11}/_{16}$–$1\,^{3}/_{8}$ inches long and $^{7}/_{16}$–$1\,^{3}/_{16}$ inches wide. Hind tracks are $^{3}/_{4}$–$1\,^{5}/_{16}$ inches long and $^{9}/_{16}$–$1\,^{3}/_{16}$ inches wide.

The voice of this animal is musical, as such voices go, for it has a variety of birdlike whistles. Commonly one hears the single sharp alarm note, and the chattering series of notes when it has taken refuge in the den.

Most of the ground squirrel species fall into the large group

Figure 29 (opposite)

a. Tracks of Uinta ground squirrel, *Spermophilus armatus* (Wyoming).
b. Foot structure of Franklin ground squirrel, *S. franklinii* (Ontario).
c. Tracks of a Wyoming ground squirrel in mud (Wyoming).
d. Droppings of Uinta ground squirrel at left (lower: a less common form, from more succulent food) and the larger rock squirrel, *S. variegatus,* at right.
e. Bounding pattern in snow, Uinta ground squirrel.
f. Walking pattern on mud, Uinta ground squirrel.

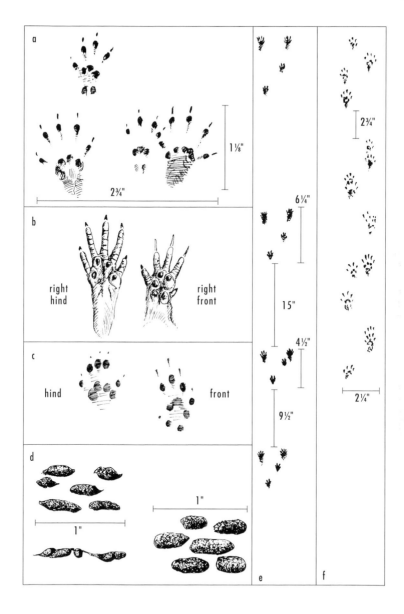

Figure 29. Ground squirrel sign

that may be referred to as gray and are often mottled or finely spotted in various ways, ranging in size from the smaller ones, such as those of the Wyoming ground squirrel, to the medium species such as the Uinta ground squirrel, S. *armatus*, to the large Arctic ground squirrel, S. *parryii* of Alaska. Arctic ground squirrel dens are 3½–5 inches in diameter and occasionally more.

These many kinds of ground squirrels all have some form of sharp whistle and trills. Those of Alaska, the Arctic ground squirrels, have a sharp double note, from which the Eskimos have given them the name *Sik-sik*. In one charming little Eskimo story and song, the ground squirrel dashes into its burrow to escape the raven, and calls sharply, *Sit-it!*

There is one species often mistaken for a chipmunk. These are the golden-mantled ground squirrels, S. *lateralis*, richly colored, and with a broad light stripe on each side bordered with black. They live in the Rocky Mountain, Cascade, and Sierra regions of the West, extending up into the mountains of British Columbia and Alberta. They prefer the rockslides, where they find convenient shelter among the many crevices, and are found in various other rocky places. Unlike other ground squirrels, golden-mantled ground squirrels are equally at home in open woods. I have not discovered any striking call notes or whistling among this group.

There is another little two-striped ground squirrel, the white-tailed antelope squirrel, A. *leucurus;* it and related species are generally known as "antelope squirrels." They are mostly gray in overall color, which is suitable to their desert environment in the arid regions of the Southwest and northern Mexico. They have the habit of scurrying along with their short tail curled up tight over their back and, since in most species the underside of the tail is pure white, it flashes bright like the white rump patch of an antelope. Front tracks measure ⅝–¹⁵⁄₁₆ inch long and ⅜–⅝ inch wide. Hind tracks are ¹¹⁄₁₆–1 inch long and ⁹⁄₁₆–¹³⁄₁₆ inch wide.

Ground squirrel tracks and trails have some characteristics that help to distinguish them from those of the tree squirrels. First, except in the more southern parts of the country, ground squirrels go into hibernation. So if in snowy country in winter you come upon tracks, they will not be those of ground squirrels. However, you may see their tracks in unusually early snowfalls in autumn, especially in the Far North; and the ground squirrels come out of hibernation early in spring while snowdrifts are still lingering.

Second, tree squirrels are found in woodlands; and whereas ground squirrels may also be found near trees in many places, they generally occupy plains, desert areas, and open country where there are no tree squirrels. Furthermore, in the Atlantic Coast states and as far west as western Ohio, Indiana, Missouri, eastern

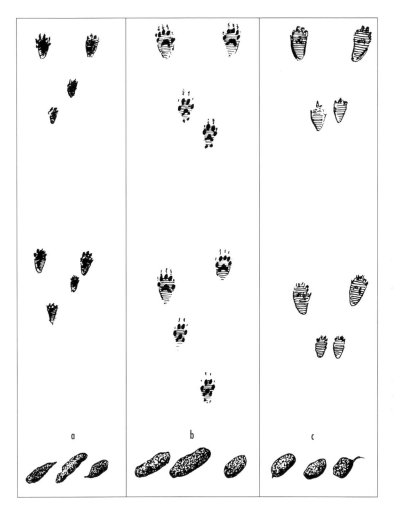

Figure 30. Tracks and droppings of ground squirrels and gray squirrel; droppings about ⅔ natural size

a. Uinta ground squirrel, *Spermophilus armatus,* tracks in snow, and droppings.

b. Rock squirrel, *S. variegatus,* tracks in mud, and droppings.

c. Eastern gray squirrel, *Sciurus carolinensis,* tracks in snow, and droppings.

Oklahoma, and eastern Texas, there are no ground squirrels.

Third, ground squirrel claws are longer and straighter than those of tree squirrels, a fact which may or may not be evident in the tracks that you find. Consider also the symmetry of the front tracks, as those of ground squirrels are far more asymmetrical than those of tree squirrels. The track patterns are on the average different. In the tracks of the tree-climbing squirrels, the pattern tends to be more square, with the two front tracks usually parallel; the pattern of the ground squirrel tracks tends to be elongated, with the forefeet tracks more in line, one behind the other. Refer to figure 30 and note the position of the front feet, more or less one behind the other, of the ground squirrel tracks in a and b, and the more parallel position in the tracks of the gray squirrel in c.

As we find in so many cases, however, these distinctions are general and do not always hold true. Very often you will find red squirrel tracks elongated in their pattern much like those of the ground squirrel.

In the samples so far collected, it appears that the droppings of the gray squirrel (and even more so those of the tassel-eared squirrels) are shorter than those of the ground squirrels (see figure 30).

The many kinds of ground squirrels vary in size, and of course there are minor differences in their tracks. Only five species are represented in figures 29 and 30: the larger Arctic ground (S. *parryii*) and rock (S. *variegatus*) squirrels, the medium-size Uinta (S. *armatus*) and Wyoming (S. *elegans*) ground squirrels, and the small thirteen-lined ground squirrel (S. *tridecemlineatus*). Note the usual variations in track pattern that we find in most mammals. The straddle is 2¼–3¼ inches in the trails of the smaller ground squirrels, and over 4 inches in that of the larger ones (see figure 30, b).

PRAIRIE DOGS

The prairie dogs of the western plains region, *Cynomys* sp., ranged from the Dakotas south through Oklahoma and western Texas to northernmost Mexico, and westward to the Rocky Mountain region. Prairie dogs have been poisoned off most of their original home, but they are still found in some national parks and refuges and on public domain where efforts to protect them are in place. A few of the places where these animals may still be observed, according to reports, are the Theodore Roosevelt National Memorial Park in North Dakota, Badlands National Park in South Dakota, the Black Hills region of South Dakota and Wyoming, and the Wichita Mountains National Wildlife Refuge in Oklahoma. Small colonies are still scattered

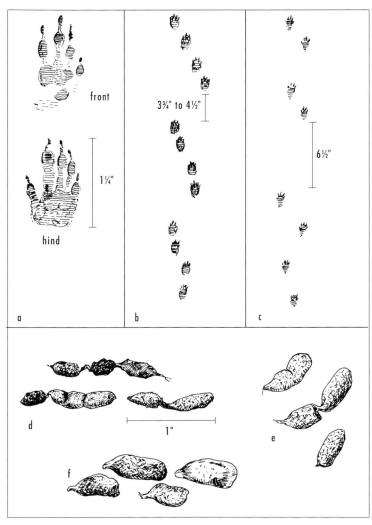

Figure 31. Prairie dog sign

a. Tracks of black-tailed prairie dog (North Dakota).

b. Typical loping pattern (North Dakota).   c. Fast gallop (North Dakota).

d and e. Black-tailed prairie dog scats, one showing less common type in which
    pellets are connected (d, North Dakota; e, Wichita Mountains, Oklahoma).

f. White-tailed prairie dog scats (southern Wyoming).

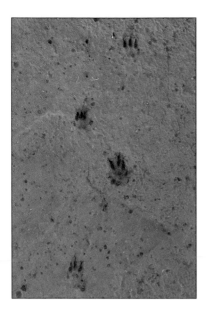

Galloping tracks of a black-tailed prairie dog in southern Colorado.

here and there in other western localities.

This is a sociable animal preferring to live in colonies or "towns," which are easily recognized by the scant vegetation and many burrows, most encircled by a raised earthen rim, in the form of a small crater.

There are several species of prairie dogs, but their tracks and other sign do not differ significantly. The tracks and gaits shown in figure 31 apply to the group as a whole; the droppings in d, e, and f show the variations. Notice that, as in the case of the marmot and some other rodents, with certain kinds of feed scats are produced in which the pellets tend to be connected, like a string of beads. In such instances the diameter or general size of the pellets is small. Figure 31, e and f, show the more common types. Front tracks of the black-tailed prairie dog measure 1 ¼–1 ⅞ inches long and 1–1 ½ inches wide. Hind tracks are 1 ⅜–2 ¼ inches long and 1–1 ⁷⁄₁₆ inches wide.

Supposedly the prairie dog got its name from the so-called bark, or cry of alarm, as it stands sentinel-like at its burrow. It flips its tail with each call. Then there is the peculiar performance when it rises up and throws its hands in the air, so to speak, as it produces a high-pitched, smooth note. It is next to impossible to describe such notes so that they can be recognized. In addition, there are various high-pitched chattering sounds.

Prairie dogs apparently do not hibernate as thoroughly as other ground squirrels. Though they remain inactive in their burrows for long periods in winter, they may be seen occasionally during those cold months.

The tracks and other signs of these three squirrels, all of the genus *Sciurus,* are similar enough to be discussed as a group. The eastern gray squirrel, *S. carolinensis,* is found in the eastern half of the United States, as far west as the eastern Dakotas, central

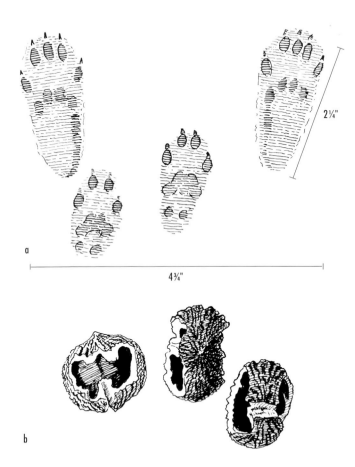

Figure 32. Gray squirrel sign
a. Tracks in ½ in. of wet snow, about ⅔ natural size (Washington, D.C.).
b. Black walnuts opened by gray squirrel (New York).

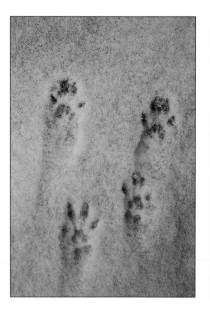

Typical bounding tracks of an eastern gray squirrel in Brattleboro, Vermont.

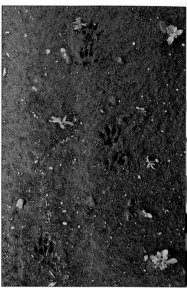

The walking trail of an eastern fox squirrel in North Dakota. This species walks often.

Kansas, and Texas, and was also released in several cities along the west coast. The larger western gray squirrel, *Sciurus griseus*, inhabits Washington, Oregon, and California. And there is the Arizona gray squirrel, *S. arizonensis*, of Arizona.

The eastern fox squirrel, *S. niger*, so much like the gray in size and general appearance, extends just a little farther west. The beautiful Abert's, or tassel-eared, squirrel, *S. aberti*, is found only in northern Arizona, western New Mexico, southwestern Colorado, and northern Mexico. In the Grand Canyon country, it is known as the Kaibab squirrel.

Figure 33 (opposite)
a. Tracks of Abert's (tassel-eared) squirrel, in wet snow.
b. Tracks of Abert's squirrel in wet snow (Grand Canyon National Park).
c. Track pattern of bounding gray squirrel (Washington, D.C.).
d. Track pattern of bounding Abert's squirrel, in snow.
e. Droppings of Abert's squirrel, about natural size.
f. Droppings of gray squirrel, about natural size (Washington, D.C.).

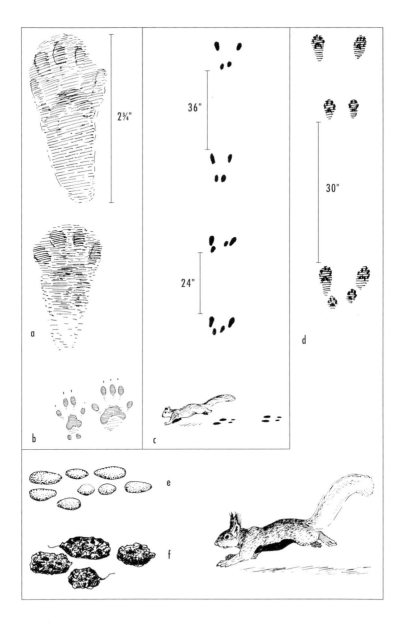

Figure 33. Tracks and scat of tassel-eared and gray squirrels

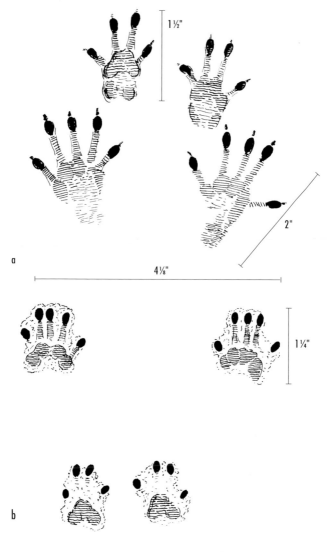

Figure 34. Fox squirrel tracks

a. Tracks in mud, about ⅔ natural size (Michigan).
b. Tracks in deeper mud, about ⅔ natural size, showing characteristic bounding pattern (Oklahoma).

The tracks and track patterns of this group of squirrels are similar to those of the red squirrel (see page 85), on a larger scale. Figure 34, a, shows Michigan fox squirrel tracks in mud; b shows Oklahoma fox squirrel tracks in deeper mud, apparently with the toes in a more cramped position. Note that the heel of the hind foot does not necessarily leave a mark. In snow, however, the mark of the entire foot is more likely to be made, as shown by the Abert's (Kaibab) and gray squirrel tracks in figures 33, a, and 32, a. In these figures there are variations in track patterns that actually are common to all species of these larger squirrels. It is important to avoid ascribing to any one species the particular pattern that happens to be illustrated for it here.

The width of trail, or straddle, varies from about 3¾ to almost 7 inches, and the leaps may be as much as 36 inches or more. Front tracks of the eastern gray squirrel measure 1¼–1¾ inches long and ⅝–1⅛ inches wide. Hind tracks are 1¼–3¼ inches long and 1–1½ inches wide. Western gray squirrels are larger still.

The droppings of this group of squirrels appear to be proportionately shorter than those of the red squirrel, and this is particularly true of those at hand from the tassel-eared squirrel. They often appear like miniature deer scats.

All of these squirrels have a varied diet of nuts, seeds, berries, mushrooms, buds, and bark. They are fond of the fruit of the elm, and of corn; and they open apples for the seeds. The storage habits of the gray and fox squirrels differ from those of the red squirrel in that their food is not hoarded in one place but buried singly here and there in the ground. They nip the buds of various trees, such as maple, elm, basswood, willow, and oak, and take the terminal tips of conifer seedlings. Like the red squirrel, they eat the bark of various trees and may remove large sections of bark from branches in the canopy. Apparently they seek the sap and the cambium layer right under the bark.

In the eastern states, as well as on the Pacific Coast, gray squirrels or fox squirrels will eat the bark of young conifers, often girdling the stems. Such gnawings on tree trunks or limbs are frequently prominent. Again, it would be difficult to distinguish this work from that of the red squirrel, or even the porcupine. The porcupine tooth marks, if they show, are larger, and if the work is recent there should be some porcupine droppings on the ground beneath the tree.

This behavior is most easily observed in the Abert's squirrel, which has a penchant for the bark of conifer twigs, notably that of the ponderosa pine. Abert's squirrels nip twigs and then completely remove the bark layer before dropping them to the ground. The forest floor surrounding trees where squirrels are feeding will be littered with such discarded twigs, which are reminiscent of discarded corn cobs. These squirrels will also gnaw bones and antlers, as several other rodents do. The voice of these large squirrels is quite different from that of the red squirrel, being deeper in tone and, in my experience, not as diversified. It is usually expressed in words as *qua-qua-qua-qua,* with some variations.

The gray, fox, and Abert's squirrels all have nests in hollow trees, as well as outside nests, just as the red squirrel does, but they are much more prone to use leaves and twigs rather than the grass and shredded bark used by the red squirrel. There may be several nests in a tree, some of them dummies without a cavity. Compare figures 35 and 38.

Figure 35. Fox squirrel with tree nest

RED AND DOUGLAS'S SQUIRRELS

The red squirrel, also called pine squirrel in its western forms, occupies much of the forested area of North America. The familiar red squirrel, *Tamiasciurus hudsonicus,* occupies the transcontinental coniferous forests from Alaska and Canada southward; in the eastern half of the United States it extends to South Carolina and in the western states it follows the Rocky Mountains down to Arizona and New Mexico. There is another species, the colorful Douglas's squirrel or chickaree, *T. douglasii,* on the Pacific Coast from British Columbia down into California. It has orange underparts instead of white or whitish, but in behavior and signs it is nearly identical to the red squirrel.

Tracks of these squirrels, leading from tree to tree, are common enough in winter snow, often appearing as a group of trails where the squirrel has run back and forth repeatedly. The squirrel's home range is small; therefore you will find its trails localized within the radius of a small group of trees. Whatever the species, or whether you call it the red squirrel, chickaree, or pine squirrel,

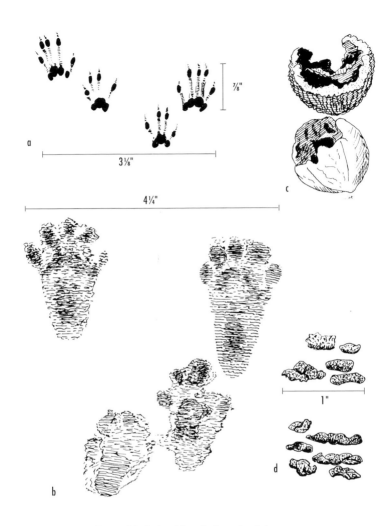

Figure 36. Red and Douglas's squirrel sign

a. Tracks of Douglas's squirrel in mud, about ⅔ natural size (Olympic Mountains, Washington).

b. Tracks of red squirrel in snow, about ⅔ natural size (Wyoming).

c. Black walnut and hickory nuts opened by a red squirrel which discovered this unaccustomed food in a cellar (Wyoming).

d. Droppings (upper, from Wyoming; lower, from Minnesota).

Bounding tracks of a red squirrel along a riverbank in western Massachusetts.

the tracks are practically the same, and the same vivacious spirit activates the animal.

Figure 36, a, shows the tracks of the Douglas's squirrel of the Olympic Mountains, clearly defined in mud. In this case the heels did not touch the ground. In b are shown the tracks of a red squirrel, in snow, with the heel marks of the hind feet showing. These two illustrate how much larger snow tracks are than mud tracks. Front tracks measure 1–1¼ inches long and ⁹⁄₁₆–1 inch wide. Hind tracks are ⅞–2¼ inches long and ⅝–1⅛ inches wide.

Figure 37 illustrates the appearance of the bounding track pattern in various depths of snow and reveals the diversity of shape. Note the variations in the position of hind and front feet — sometimes one foot forward sometimes the other, though generally the hind feet are in front. One could almost say that in the red squirrel track pattern, being made by a tree-climbing rodent, the front feet are parallel, forming a "square-sided" group pattern, in contrast with the elongated pattern of the ground squirrel, which generally puts one forefoot in front of the other when bounding. But you will notice in the trails pictured here that many times the red squirrel also puts one forefoot in front of the other. We can only say that the red squirrel tends to keep the forefeet parallel when running, and one should glance at the trail as a whole to catch this tendency.

Where toe marks show, it should be remembered that there are five toes on the hind feet, four toes on the front feet. And the leaps vary from about 4 to 30 inches. The width of the track group, or straddle, varies from 2⅞ to 4½ inches, even appearing a little more in very loose snow.

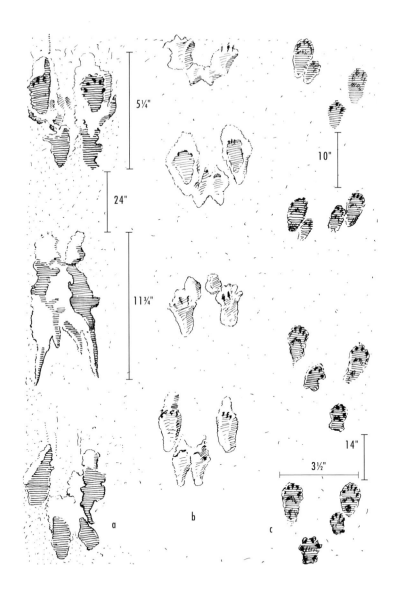

Figure 37. Douglas's squirrel bounding trails
showing variations due to snow conditions

The red squirrel makes three kinds of nests—the outside nest, as shown in figure 38, a; the one in a hollow tree, such as the one from Minnesota in b; and those created in underground burrows. The outside nest, about a foot in diameter, is more common in coniferous forest, where hollow trees are less available. They are roughly globular, with entrance in the side, and may be built of grass or fine twigs, and lined with shredded bark. I found one built almost entirely of caribou hair, held together with a mixture of grass and twigs. Sometimes two or three of these nests are built in one tree. One winter a red squirrel built a nest in a box inside our cellar house. A weasel built a nest on a shelf in the same house, and the two spent the winter under the same roof.

The hollow tree dens (often in old flicker holes) are more common in deciduous forest, where such cavities are more available. In the eastern states I have found the red squirrel taking advantage of the cavity that has rotted out where a limb has broken off. In living trees, occupied cavities are very apparent, as brightly colored bite marks will ring the entrance.

Red squirrels also have holes in the ground, especially in the mass of cone scales, or middens, where cones are stored. They will tunnel into the snow to reach the stored food, and are known to nest in middens and cone caches in winter months.

Once in a heavy coniferous forest of Oregon I heard a loud thump as if someone had struck a log with a club. Again and again I heard the sound, and at times, when several blows seemed to come in rapid succession, it suggested the clashing of deer antlers and I began to look for the deer. When I traced the sounds to a large sugar pine, the mystery was solved. A huge cone came hurtling down from the top, whacking limbs here and there, finally landing on the ground with a great thump. Up near the top was a squirrel, busy cutting loose one cone after another for its nut harvest.

The autumn is the harvest time for squirrels, east, west, and north. East or west, you may find them carrying mushrooms into the trees, placing them on limbs to dry. They store up acorns and other nuts. Cones of spruce or pine are cut loose and gathered on the ground, to be tucked away in odd corners for future use, especially in the midden heaps of old cone scales, under logs, or in tree hollows. In Alaska certain berries, such as those of the viburnum, are stored. In the Middle West and eastern states these squirrels are fond of the seeds of basswood and box elder.

When spruce and lodgepole pine cones are severed, the twigs containing the cones are cut off, and in such places you will find the ground underneath strewn with spruce or pine twigs. The porcupine will do the same, but if it is porcupine work you will

Figure 38. Red squirrel nests
a. Outside nest in spruce tree (Alaska).
b. Hollow tree den (Minnesota).

generally find scattered porcupine droppings on the ground, and the nipped twigs will measure much longer.

Red squirrels will sometimes nip off the buds of young spruce trees and gnaw the bark and twigs of some trees. John Pearce, of the U.S. Fish and Wildlife Service, pointed out that red squirrels will gnaw fresh blister-rust cankers on white pine. Some of these gnawings resemble those of the porcupine. However, if tooth marks show, those made by the red squirrel are about as fine as mouse work, compared with the coarse marks left by the porcupine.

The squirrel midden, sometimes covering several square yards of ground and consisting of a great accumulation of cone scales over a period of years, marks the squirrel's home ground. There is the place where cones have been stored and eaten. Also, you will find little piles of scales on a log or stump or hillock where the squirrel has feasted on cones.

A Wyoming red squirrel found our store of Christmas nuts, black walnut and hickory, sent to us by eastern friends. These nuts were of course unfamiliar to this individual squirrel, but it recognized the value of the nuts and gnawed them open, as shown in figure 36, c. Contrast this operation with the opening of such nuts by gray squirrels and flying squirrels, illustrated in figures 32 and 39.

One day in the Snake River bottomlands in Wyoming, I noticed many green leaves on the ground under a large cottonwood. While I puzzled over this, several more leaves came fluttering down. Then I spied a red squirrel high in the tree, nibbling loose some more. The answer became clear. At the base of each leaf was a large insect gall. The squirrel was busy opening the galls for the larvae inside and, from evidence of the number of leaves on the ground, a considerable percentage of the insect larvae were eaten.

The voice of the Douglas's squirrel, or chickaree, is distinctive. One cannot mistake the scolding notes and the varied calls of this most dynamic little creature. There may be the prolonged chatter, then higher-pitched notes, in great variety and often quite explosive in effect. It may be a *tsik, tsik, tsik, chrrrrrrrr—siew, siew, siew, siew,* if one may presume to put into human syllables one version of red squirrel language. Actually, the tirade may be a series of coughs and hiccups and high-pitched notes in a combination impossible to write, but they are easily recognized once they become familiar.

FLYING SQUIRRELS

A flying animal may be expected to leave rather sketchy trails, but so does the red squirrel, since its trails are largely from tree to tree. In fact, often one must look sharp to distinguish between

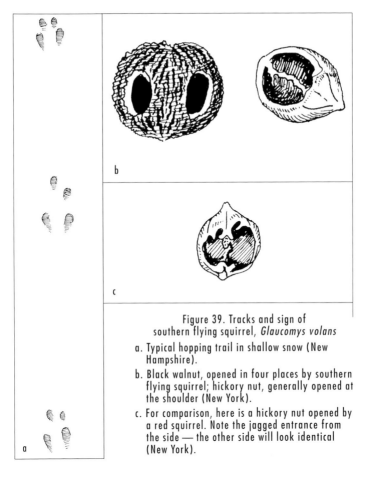

Figure 39. Tracks and sign of southern flying squirrel, *Glaucomys volans*

a. Typical hopping trail in shallow snow (New Hampshire).

b. Black walnut, opened in four places by southern flying squirrel; hickory nut, generally opened at the shoulder (New York).

c. For comparison, here is a hickory nut opened by a red squirrel. Note the jagged entrance from the side — the other side will look identical (New York).

the tracks of the larger northern flying squirrel, *Glaucomys sabrinus*, and those of red squirrels. Generally speaking, the tracks of the flying squirrel are smaller; but snow tracks vary so much in size, depending on the condition of the snow, that positive identification is often difficult. Moreover, in deep snow the red squirrel may show drag marks of the feet somewhat like those of the northern flying squirrel.

When you find a "sitzmark," or landing spot, in an open area from which tracks lead off, you know that the flying squirrel "set himself down" there. See figures 40, a and c, and 41, b. In 41, b,

 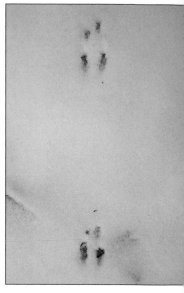

The bounding trail, moving up from bottom, of a northern flying squirrel in southern Vermont.

The trail, also moving up, of a southern flying squirrel in southern New Hampshire.

the drag marks extended for some 50 inches (including the tail mark). In such landings the body leaves a definite gouge in the snow.

The trails of southern flying squirrels, *G. volans,* are smaller and more distinct. They are more the size of chipmunk trails than of red squirrels. Unlike all other rodents of similar size, southern flying squirrels tend to bound in such a way that the front tracks remain in front of the hind tracks. Refer to figure 39. Often naturalists mistake their trails for those of chipmunks and assume the animal is traveling in the opposite direction!

Figure 40 (opposite)

a. A landing mark, in snow, with trail leading off in bounds of 14, 21, and 20 in. (all tracks in this figure from Wyoming).

b. Snow trail with bounds of 29, 21, and 11 in. In one spot the squirrel slid.

c. Another landing mark, 9½ in. long, with irregular track pattern.

d. More flying squirrel tracks, in slow bound.

e, g. Scats, about natural size (from Wyoming and New York, respectively).

f. Scats of southern flying squirrel (Washington, D.C.).

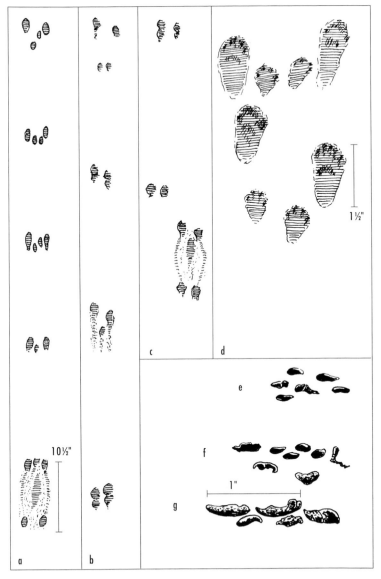

Figure 40. Tracks and scats of northern flying squirrel, *Glaucomys sabrinus*

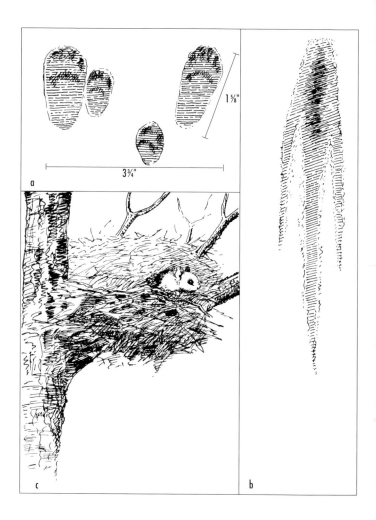

Figure 41. Northern flying squirrel sign
a. Tracks in snow (Wyoming).
b. Landing mark in snow, length 50 in., where the animal slid on the surface.
c. A renovated bird's nest (Wyoming).

Figure 40, e, f, and g, shows droppings from Wyoming, Washington, D.C., and New York. There is probably much variation in size, depending on food, but on the whole the scats are smaller than those of the red squirrel. Figure 39, b, represents a black walnut and hickory nut opened by a southern flying squirrel, with the openings on the side.

Both outside nests and hollows in trees are used. On the bank of the Red River in Minnesota I came upon a mother flying squirrel with a family of young in a hollow in the side of a tree, less than 5 feet from the ground. Other nests are in woodpecker holes or similar cavities high in the trees.

One early spring day in Jackson Hole, Wyoming, I climbed a fir tree to examine an old bird's nest, possibly built by Steller's jays. As I came within a couple of feet of the nest and had just noticed that it was domed over with added material, it seemed to explode when several flying squirrels burst out of it and sailed away in different directions. I watched one plane far down the mountain slope. It pays to examine old birds' nests. Flying squirrels will also use other squirrels' nests, and attics of houses.

One moonlit night in Minnesota when two friends and I were camping out, tentless, we noticed shadowy figures sailing above us from tree to tree. Then we heard the quiet tick of the animals, which proved to be flying squirrels, landing on the trunks of trees. Next we heard the rustlings and gnawing as they gathered the seeds of the ash trees. Faint sounds up in the trees at night may call your attention to the aerial activities of these nocturnal squirrels.

POCKET GOPHERS

Once when I was out on a mountainside with a troop of Boy Scouts to help them interpret what they saw, we stopped at a fresh mound of earth.

"What has happened here, boys? Who did that?"

"Mole," someone declared. "Mole, mole," was repeated by one after another.

So I had to explain, "There isn't a mole in this part of the Rocky Mountains. Ever hear about the pocket gopher?"

The boys reflected the confusion prevalent everywhere about these two unrelated animals. We examined the mound of earth. "See that little hollow there, with the earth plug in it? That's where the animal finished digging. When he wanted to dig a new burrow here he pushed out the dirt, and as the pile grew, he came out and pushed it over pretty much to one side. Then, when he finished his excavating down there, he pushed up enough dirt to plug the entrance. That's the little dirt cap you see there at one

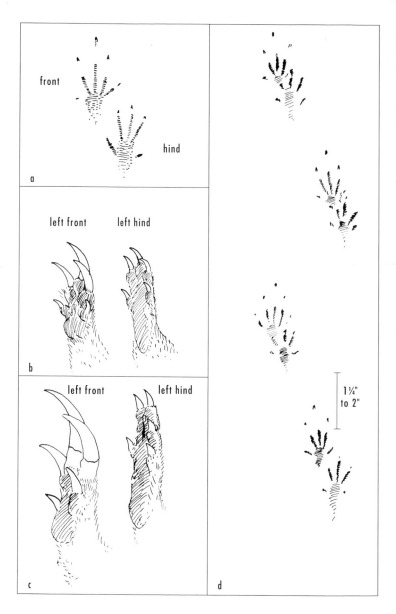

Figure 42. Pocket gopher

(Left) *The walking trail of a northern pocket gopher in Yellowstone National Park.* (Above) *The trail castings of northern pocket gophers in the Colorado Rockies.*

side of the mound. If that had been a mole, he would simply have pushed that earth upward from below, without shoving it to one side, and there would be no earth cap like the one we see here." (See figure 43.)

We walked on a little and stopped again. "Now, can you explain that?"

No, they couldn't. There were some heavy ropes of earth lying on the ground, earth cores exposed when the snow had thawed away in spring. Pocket gophers continue their digging in winter, too, and shove the earth into snow tunnels. These ropes of earth were the mud casts of snow tunnels, about 2 inches in diameter, of the northern pocket gophers.

Pocket gopher tracks may be encountered at any time of year, but seem to be most common in spring when high water levels, gathering nesting materials, and feeding might bring them above ground to travel. At any time of year they may travel to usurp an

Figure 42 (opposite)
a. Tracks of northern pocket gopher, *Thomomys talpoides*, in mud; natural size.
b. Left feet of *T. talpoides*, natural size.
c. Left feet of the much larger plains pocket gopher, *Geomys bursarius*, from Minnesota, natural size. Note long front claws used for digging.
d. Walking track pattern of *T. talpoides*, of Wyoming, in mud. About ⅔ natural size; straddle 1½–2 in.

abandoned tunnel network of another pocket gopher. Figure 42, a and d, illustrates the tracks in mud. Note that there are five toes in both front and hind tracks, when they all show, and that the front claw marks are far out in front. Note also the much greater size of the feet of the plains pocket gopher (*Geomys bursarius*), shown in figure 42, c, than those of *Thomomys*, in b.

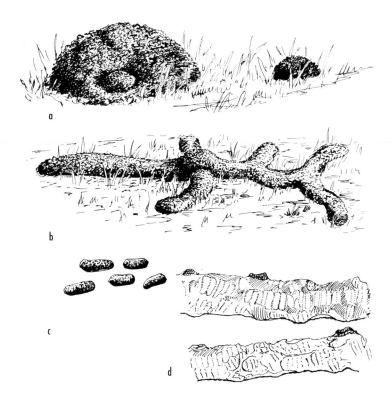

Figure 43. Pocket gopher sign

a. A large mound, showing earth plug near one edge, and a small mound that itself constitutes the plug.

b. Earth cores that had been pushed into snow tunnels in winter.

c. Pocket gopher droppings (Wyoming).

d. Aspen limbs gnawed by northern pocket gophers, individual tooth marks about ⅟₁₆ in. wide.

The front track of the northern pocket gopher, *T. talpoides,* measures ⅞–1 5⁄16 inches long and 7⁄16–½ inch wide. Hind tracks are ⅝–⅞ inch long and 7⁄16–1 1⁄16 inch wide. The front track of the plains pocket gopher, *G. bursarius,* measures 1½–1⅝ inches long and ⅞–1 inch wide. Hind tracks are 1⅛–1¼ inches long and 1⅛–1¼ inches wide.

One spring day in Yellowstone Park I followed the tracks of a grizzly on intermittent patches of snow and found an excavation he had made in an open meadow. The story was plain. His keen nose had located a pocket gopher cache of roots a few inches underground, and he had feasted with satisfaction. Some of the roots, half a dozen kinds, lay scattered about. I was studying the bears at the time and so took the roots. Since I did not know what plants they came from, I took them into my camp and planted them. When the plants grew I could list them as the food of pocket gophers for the long winter under the snow, as well as of the robber grizzly.

Under the winter snow the pocket gophers still forage over the ground within snow tunnels, and when the snow thaws in the spring you will find fallen limbs and bases of small trees and bushes gnawed by these rodents. There will be mud cores nearby, and the limbs will show that the teeth have gouged deeply into the wood, leaving a very ragged surface. The individual tooth marks are about 1⁄16 inch wide, a little wider than those of mice.

It should be explained that there are three genera of pocket gophers: *Thomomys,* with its multitude of forms, large and small, occupies the western states and Canada, and part of Mexico; *Geomys* is the one in the Great Plains region and southeastern states; and *Cratogeomys* is found in the Southwest and Mexico.

POCKET MICE

The pocket mice of the genus *Perognathus* are distributed over the western half of the United States and Mexico, usually in the more arid plains sections. The one whose tracks are pictured here is *P. longimembris.* Other species have a variety of hair structures, and they vary in size. Clean tracks of pocket mice

Bounding tracks of a little pocket mouse in southern California.

are distinct among mice. All feet show long toes and significant claws. The hind feet have five toes like other mice, but their inner-most toe is greatly reduced in size and may not appear in tracks (see figure 44).

Their trails are similar in pattern to those of the white-footed mice, but are smaller, the straddle being between ¾ and 1 ¼ inches (see figure 44, a). Note also that their front feet tend to overlap, rather than registering independently as in other mice. Their bounds vary from 2 to 5 inches, and tail marks are common when pocket mice are foraging or moving slowly. The droppings are tiny, black, and seedlike in appearance (see figure 44, b).

Although it is difficult in many cases to identify burrows, the pocket mouse burrows generally have near the entrances a mound of fine soil, much like that made by the pocket gopher, but usually smaller. Like the pocket gopher, too, they may plug branch burrows with loose soil. The mounds, unlike those of the pocket gopher, are not made in spring or early summer, except by

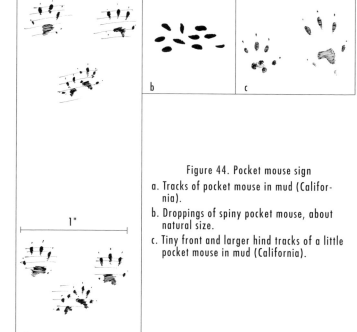

Figure 44. Pocket mouse sign

a. Tracks of pocket mouse in mud (California).

b. Droppings of spiny pocket mouse, about natural size.

c. Tiny front and larger hind tracks of a little pocket mouse in mud (California).

some species. The burrow itself, a little less than an inch in diameter and a little smaller than that of *Peromyscus* (see page 112), goes to a depth of several feet, with the nest far underground. Seed caches may be made closer to the surface.

Pocket mice may be dormant during a cold part of the winter.

KANGAROO RATS AND KANGAROO MICE

Kangaroo rats of the genus *Dipodomys* are interesting rodents and inhabit the plains and deserts of the West. There are 16 distinct kangaroo rat species, and although they share much in behaviors and habits, they vary tremendously in size. The tiny Merriam's kangaroo rat, *D. merriami,* weighs 33–54 grams, while the large desert kangaroo rat, *D. deserti,* weighs 83–148 grams. Kangaroo rat species are generally split into two large groups: those with four toes on the hind feet, and those with five. However, even those species with five toes on each hind foot often leave what appear to be four-toed tracks, as their innermost toes are greatly reduced.

As one would expect of such a "kangaroo" structure, travel is performed by bipedal hops, with the hind legs only, and when traveling at any appreciable speed the animal runs on its hind toes (figure 45, c). When shuffling along slowly to feed, all four feet are down, including the heels of the hind feet, as in figure 45, a; like foraging pocket mice, the tail drags in the sand.

Built on the same lines, but smaller, is the so-called kangaroo mouse, well named *Microdipodops*. It is relatively rare, or at least little known. It looks much like its larger relative but has no brush at the end of its tail. It leaps in the same kangaroo manner, leaving the twin tracks shown in figure 45, d.

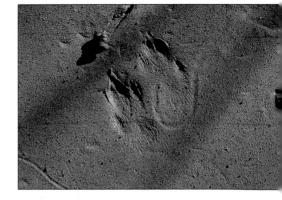

Perfect hind tracks of an agile kangaroo rat moving in a bipedal hop in southern California.

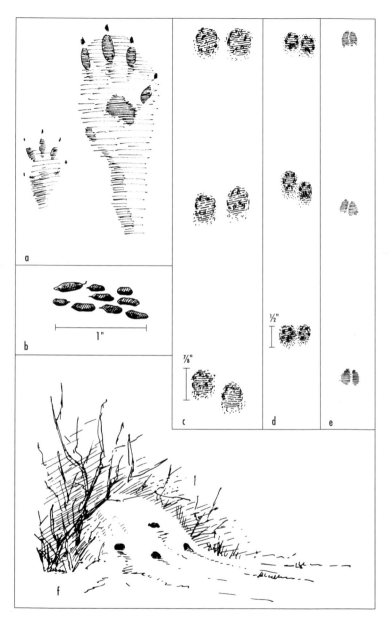

Figure 45. Kangaroo rat and kangaroo mouse sign

Both of these, when traveling slowly, will put down the front feet, and then more of the hind foot may show as the animal lands on it more solidly. In leaps, the hind heel is off the ground and tail marks do not show.

Comparative measurements are about as follows: outside width of straddle in the trail for Merriam's kangaroo rat, ⁹⁄₁₆–1½ inches; for desert kangaroo rat, 1⅜–2⅝ inches; for kangaroo mouse, about 1⅛ inches. A kangaroo rat will leap from as little as 5 or 6 inches up to 60 inches, and some species are reported to be able to leap upward of 8 feet or more.

In all kangaroo rat species the front tracks are significantly smaller than hind tracks. Front tracks for Ord's kangaroo rat, *D. ordii,* are ⅜–⁹⁄₁₆ inch long and ⁵⁄₁₆–⅜ inch wide; hind tracks are ⅜–1½ inches long and ⅜–¹⁵⁄₁₆ inch wide. Front tracks for the desert kangaroo rat, *D. deserti,* are ⅝–¾ inch long and ⅜–½ inch wide; hind tracks are 1⅜–2³⁄₁₆ inches long and ¹¹⁄₁₆–1 inch wide.

The scats of *Dipodomys* are from about ⅛ to ⁹⁄₁₆ inch long, and may be dark green or brown in color. Those of the kangaroo mouse are similar but smaller, from about ⅛ to ⅕ inch. Droppings may be found in the den tunnels, on the ground near the burrows, or where they are feeding. Desert kangaroo rats form large latrines near sticks and other features within their territories, where they may also rub their cheek glands.

A characteristic feature of desert areas consists of the mounds of sand or fine soil thrown up by kangaroo rats. Those of the largest species may be over 3 feet high and over 12 feet in diameter. These have a number of burrow entrances, 4 or 5 inches in diameter. Smaller species have smaller mounds and burrows, and some do not throw out conspicuous mounds. Usually you will find some of the burrow entrances plugged with loose earth or sand.

You will find trails leading from one mound to another, or radiating to feeding places. It should be kept in mind that the pocket mice also may throw out a small mound. Kangaroo rats enjoy a dust bath, so you will find little dusting spots, which are often scooped out slightly.

Figure 45 (opposite)

a. Tiny front and larger hind tracks of Gulf Coast kangaroo rat, in moist sand (Texas).

b. Droppings of a Nevada kangaroo rat, about natural size.

c. Tracks of kangaroo rat in sand, only the toes touching (Black Rock Desert, Nevada).

d. Tracks of kangaroo mouse, *Microdipodops,* in sand (Black Rock Desert).

e. Bipedal trail of a Merriam's kangaroo rat (Big Bend National Park, Texas).

f. Burrow mound of kangaroo rat.

There is another indication of the presence of this rodent. If you scrape or tap at the entrance to an occupied burrow, you may hear within a light thumping sound, which has been referred to as "drumming." The woodrat will also thump with its feet, and so does the skunk. The kangaroo rat shares this habit.

Around the burrows and along the paths leading from them you may find fragments of grass or other plants that have been cut.

AMERICAN BEAVER

The familiar *Castor canadensis,* which had contributed to the fur supply so importantly, is pretty well distributed over North America, wherever it has been able to find its food among forest products. It is the rodent that most obviously affects the landscape, and its works are the easiest to discover, even if you never see the footprints.

The dam and the lodge are the most obvious beaver sign. These (suggested in figure 46, e) are so obvious that for recognition purposes they need no further comment here. It should be remembered, however, that the beaver also digs burrows in banks of ponds, lakes, and rivers, with an underwater entrance. Thus "bank beavers" may have built no lodge, depending entirely on the burrows. What the beaver is after is deep enough water for winter. This is present in streams of large enough size. Ponds are built by damming the smaller streams, for the same purpose.

Beaver cuttings, too, are prominent. Fallen trees and stumps will show the characteristic gnawing techniques that felled the trees— simply an encircling cut in a deepening groove, until the tree falls. The tooth marks and the chips reveal the workman. On the bank, or in the water, you will find peeled logs and twigs, smooth and gleaming white when fresh. These are the remains of the feast on bark. Aspen, cottonwood, birch, and willow are some of the favorite food trees, though a beaver will occasionally gnaw some of the bark of a standing pine, and conifers are sometimes felled.

Figure 46 (opposite)
a. Tracks in mud. Hind track, 6–6½ in. by 4¾–5¼ in.; front, about 3 in. long (Wyoming).
b. Walking trail in mud, with strides 6–9 in. long (Wyoming).
c. Tooth marks, natural size. See also figure 194, e, f, g.
d. Scats, natural size.
e. Beaver dam, pond, and lodge, with beaver leaving scent mound in foreground.

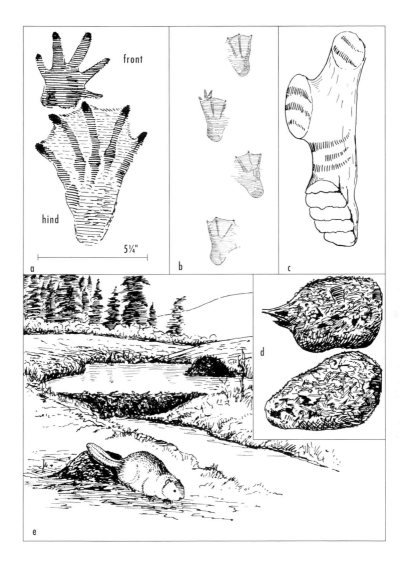

Figure 46. Beaver sign

The tooth marks of beaver are relatively broad, usually from ⅛ to ¼ inch in width, more commonly ⅛ inch. See figures 46 and 194.

The scent mound is a pile of mud scraped together, or a pile of mud, grass, and sticks, on which the beaver leaves castoreum from the glands developed for that purpose (figure 46, e). This is similar in function to the "sign posts" of some of the carnivores. Along the shore you will often see little dabs of mud with stems of sedges, apparently perfunctory efforts to establish scent piles. The more elaborate ones may be over 3 feet in diameter and upward of a foot in height.

Droppings are not often found, since they are most often deposited in water. They consist of oval pellets of coarse "sawdust." Occasionally several pellets may be temporarily connected, like beads, but being in the water, such material soon disintegrates.

Scats may be 1–1¼ inches long, and about ¾ inch in diameter, and of course will vary in size and shape (see figure 46, d).

Perfect beaver tracks are hard to find, for the tail often drags and obscures them. Nevertheless, they are distinctive because of the large webbed hind feet. Rarely do you find all five toes of the front or hind tracks. More often they appear as three- or four-toed tracks. Front tracks measure 2½–3⅞ inches long and 2¼–3½ inches wide. Hind tracks are 4¾–7 inches long and 3¼–5¼ inches wide.

The walking trail of an American beaver along a lakeshore in western Massachusetts.

There are well-beaten trails across dams from one pond to another, or overland between two bodies of water, and leading into popular feeding areas, such as corn fields. You will find drag marks on the ground where beavers have dragged limbs or lengths of small logs to the water, and you may find the telltale chips where the wood was cut. In connection with beaver ponds, too, there may be long canals, sometimes several hundred feet long, dug by the animals for the purpose of floating in logs.

In snow country, when the beaver climbs out of the water for foraging in the woods, it leaves a trough that at first may be mistaken for that of the porcupine or otter, especially at a distance. However, any fragment of a footprint should identify it.

Sometimes you find beaver cuttings or patches of bark eaten from a tree trunk at such an astonishing height above the ground as to suggest a true "giant" beaver. The mystery is solved if one is mindful of the fact that these cuttings were made from the top surface of deep winter snow, which may have held the beaver several feet above the ground.

Occasionally you will come to an area, perhaps an old pond or marshy place that used to be a pond, surrounded by ancient dead trees. You may find the remnants of a beaver dam built long ago. This is what happened. The beavers built a dam that permanently flooded the surrounding trees. These eventually died. As the pond, in the evolution of such landscape, becomes a meadow, trees will once more cover the area.

The beaver's voice is not conspicuous, but he does make noise. When excited, or alarmed, he dives and slaps the water with his flat tail, causing a loud resounding splash. Whether intentionally or not, no doubt this serves as a signal to fellow beavers. Should you approach the lodge, you can sometimes hear the chirplike sounds of young beavers inside.

RICE AND COTTON RATS

The rice rats, *Oryzomys* sp., are found in the eastern United States, from eastern Kansas and Oklahoma onward and from New Jersey south to Florida and Central America. The cotton rats, *Sigmodon* sp., are found "in greatest abundance from Mexico to Peru," but are also plentiful throughout our southern states.

The tracks of these two rat groups are quite similar, as suggested by figure 47. They are also similar to the tracks of Norway rats, although smaller. Both make runways in the vegetation that are hardly distinguishable. Both also make nests of grass or leaves, above or below ground. The front tracks of the hispid cot-

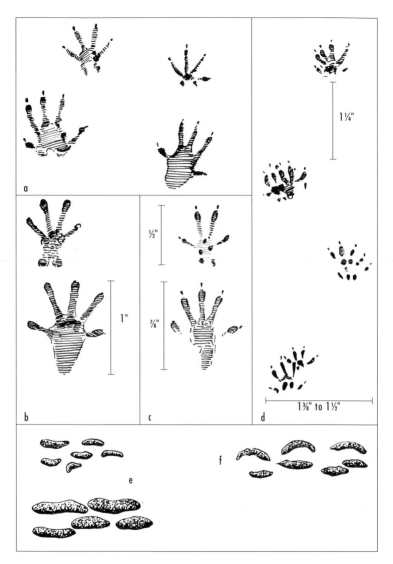

Figure 47. Tracks and droppings of cotton rat and rice rat

The trotting trail of a hispid cotton rat in southern Texas.

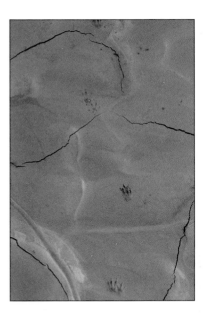

ton rat measure $\frac{7}{16}$–$\frac{5}{8}$ inch long and $\frac{7}{16}$–$\frac{5}{8}$ inch wide. Hind tracks are $\frac{9}{16}$–$\frac{7}{8}$ inch long and $\frac{9}{16}$–$\frac{11}{16}$ inch wide.

In cases like this, one must keep in mind geographic distribution of animals. When you find small rodent tracks in Montana or Maine or Canada, you can rule out both the cotton rat and rice rat. Though long, slender fingers will help rule out mice, voles, and woodrats, you will likely still need to consider potential confusion with young Norway rats.

It is interesting to note, also, the difference in size between the scats of a captive cotton rat in the zoo and those of wild rats from Texas, as shown in figure 47, e.

Figure 47 (opposite)
a. Cotton rat track pattern, when at rest; natural size (Chisos Mountains, Texas).
b. Cotton rat tracks in mud, showing heel of hind foot; natural size (National Zoological Park, Washington, D.C.).
c. Rice rat tracks, in mud; natural size (National Zoological Park).
d. Walking gait of cotton rat, in mud.
e. Cotton rat scats, natural size (upper ones from zoo animals; lower ones from Chisos Mountains).
f. Rice rat scats, from zoo animals; natural size.

Harvest mice, *Reithrodontomys* spp., are inconspicuous creatures seldom seen. They are found in certain areas throughout the United States, particularly in grassy plains or deserts, and range into Central America.

The most conspicuous evidence of the presence of these mice is the globular nest they build, approximately 3–6 inches in diameter, either on the ground or in bushes, or invisibly underground.

On one occasion in a canyon in northern Nevada I found a bird's nest in a large sage bush filled and domed over with the feathers of a flicker. Snug inside this adapted structure was a harvest mouse.

Harvest mice will also build small feeding platforms where one will find cut twigs and scats. In Nevada there were little platforms of sage twigs with leaves lodged in forks of the branches of tall sagebrush. These bushes, it was determined, were climbed by both the harvest mice and the white-footed mice.

The trails of harvest mice are most similar to those of house mice, in that they are smaller dimensions than trails left by white-footed mice species. Individual tracks must be studied for confident identification, as the front tracks are much narrower than white-footed or house mice. The hind tracks are also distinct, for the innermost toe is set farther back on the foot than in other mice. Although this is also true for pocket mice, their innermost

Figure 48. Harvest mouse tracks and droppings

a. Enlarged front and hind tracks in mud (California).

b. Droppings are of typical mouse characteristics: small, dark, and scanty in number for each deposit; vary with size of animal (drawn by Carroll B. Colby from specimens in the American Museum of Natural History, New York).

The perfect nest of a harvest mouse, constructed with the downy material from flowering coyote brush, in Santa Barbara County, California.

toes are also reduced in size. In harvest mice, each of the five toes is well developed. Front tracks measure ¼–⅜ inch long and ¼–5⁄16 inch wide. Hind tracks are ¼–½ inch long and 7⁄32–5⁄16 inch wide (see figure 48).

This dainty little mouse also has a song, which Ruth Svihla has described as "a tiny, clear, high-pitched bugling sound, very ventriloquistic in character." It would be interesting to know more about the accomplishments of these obscure animals that we so seldom see.

The tracks of a western harvest mouse in southern California.

The white-footed mouse and the deer mouse are members of a large group commonly referred to as white-footed mice, genus *Peromyscus*. This well-known and abundant group includes numerous species and subspecies. Certain species make their homes in arid regions of the Southwest, while others have adapted themselves to all habitats, from the plains to the mountaintops and from Central America into much of Canada.

The usual sign is the four-print pattern in the mud or snow, as shown in figure 49, c and d. This bounding pattern may be confused with that of the house mouse, and also resembles that of the small shrews. However, the straddle of the white-footed mouse trail is larger than that of the shrew, being 1 ½–1 ¾ inches. Sometimes the tail drags, as in figure 49, b, and in this type the front and hind feet make one elongated print in the snow on each side, suggesting certain arrangements of weasel tracks.

The front feet of white-footed mice species have five toes, yet you must look carefully to see the innermost toe, which registers only as a dot in clean tracks, if at all. The hind feet have five distinct toes. White-footed mice tracks are tiny versions of woodrat tracks. Their digits appear bulbous in tracks, rather than the long slender toes seen in the tracks of voles, shrews, and jumping mice. Front tracks measure ¼–⁷⁄₁₆ inch long and ⁵⁄₁₆–½ inch wide. Hind tracks are ¼–⁹⁄₁₆ inch long and ⁵⁄₁₆–½ inch wide.

White-footed mice are primarily seedeaters and you may find as much as a quart of seeds stored for winter use in any convenient hole or hollow or cranny. They open large fruit pits, such as the wild cherry pits shown in figure 50, d, for the kernels inside. The hazelnut pictured in c was one that the mice found in the cupboard, uncommon food for the

The walking tracks of a white-footed mouse in southern New Hampshire.

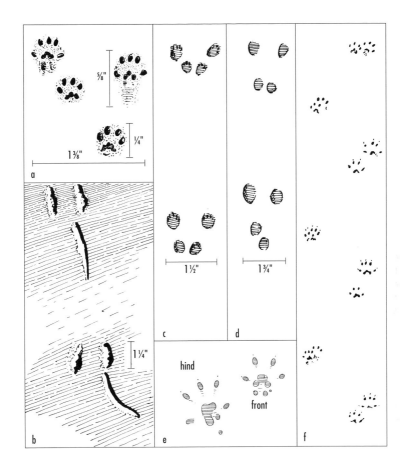

Figure 49. White-footed mouse tracks, from Wyoming

a. Bounding tracks in dust, about natural size. Front track (4-toed, as the innermost 5th toe is greatly reduced in size) ¼ in. long and wide.
b. In light snow, with tail marks. Hind and front feet make one elongated print, about 1¼ in. long. Leaps measured 3–9 in.
c. On very light layer of snow. Bounds 5–8 in.
d. On firm snow. Leaps 2–3 in.
e. Front and hind tracks in mud (New Hampshire).
f. Walk in wet mud. Individual tracks about ⁵⁄₁₆ in. wide.

mice of northern Wyoming. In eastern forests the mice open the basswood seeds, and you will find many of the shells with a hole in them.

These mice do not cut runways through the grass, as do the voles, but live among logs and stumps. They will climb up on tree trunks and into bushes; they live in holes in plowed furrows, among rocks, and on ocean beaches. In fact, the white-footed mouse as a genus is extremely adaptable and finds a congenial home almost anywhere, though many of its species have chosen special habitats.

The nests vary in size from a few inches to nearly a foot in diameter, and their location varies greatly—burrows, logs, holes in trees, even bushes. Often a white-footed mouse will roof over an abandoned bird's nest and live in it (figure 50, a). It is sometimes difficult to distinguish signs of harvest mice from those of *Peromyscus*.

Mr. R. DeWitt Ivey has given us a published account of the habits of three species of white-footed mice on the east coast of Florida. He found that the so-called golden mouse, *Ochrotomys nuttalli,* lives in loose communities of three or four nests. These nests, sometimes found at heights up to 15 feet, are usually 4–6 feet above ground, placed in Spanish moss, vines, or other vegetation. They are globular, 3–6 inches in diameter, with an opening about 1 inch in diameter.

The cotton mouse, *P. gossypinus,* makes globular nests in hollow logs and stumps and on the ground. The smaller oldfield mouse or beach mouse, *P. polionotus,* builds its nests within ground burrows. Mr. Ivey warns us, however, that "burrows are not reliable as evidence of the presence of [oldfield] mice except when excavated, since they may be readily confused with the burrows of the sand crabs."

Everywhere you go you will find these little mice, adapted in color and habits to their special surroundings. We find them abundantly among the forests and fields of New England and all the eastern and midwestern states; the little white ones on the Florida beaches; other species have adapted to the mountains and plains of the West—the pale-colored ones in the desert, the big, dark-colored ones in the Pacific Coast rain forests. But everywhere, from the subtropics to the high western mountains, the little foursquare track pattern reveals them.

Generally speaking, all mice squeak in the manner familiar to us. Some mice sing and, according to some opinions, both the house mouse and the white-footed mouse may have a song too high in pitch for the human ear, but descending in pitch on rare occasions to become audible. The series of notes, when heard, are described as a birdlike trill that can be heard only at a distance of a few feet.

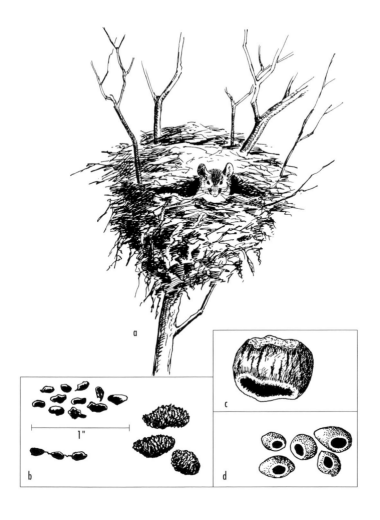

Figure 50. White-footed mouse sign

a. Old bird nest roofed over by a *Peromyscus* sp.

b. Droppings, natural size; at right, three samples much enlarged (Wyoming).

c. Hazelnut opened by white-footed mouse (in kitchen cupboard). This Wyoming mouse would not find wild hazelnuts.

d. Wild cherry pits opened by white-footed mouse (New York).

I recall a number of times when I lay in my sleeping bag in the woods and could hear the tiny patter of mice across my bed. When a mouse came very close to my ear I could hear little vocal sounds, a rapid series almost like a chatter, but very faint and less forceful than that term implies.

GRASSHOPPER MICE

Mice of the genus *Onychomys* are largely insectivorous and carnivorous, and are found in the western United States and adjacent parts of Canada and Mexico. The tracks of the grasshopper mouse have seldom been observed or recorded, or at least positively identified in the field. The tracks in figure 51 were roughly sketched from a photograph of tracks of a captive animal. The scats also shown here are from a captive animal. Note the difference in size between the moist fresh ones and the older dried ones.

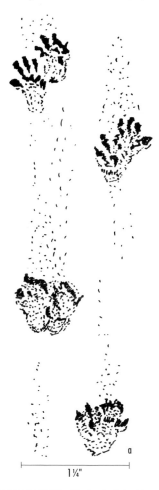

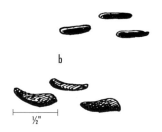

Figure 51. Tracks and scats of grasshopper mouse

a. Tracks of captive mouse (drawn from a photograph by Vernon Bailey, courtesy of the U.S. Fish and Wildlife Service).

b. Droppings, about natural size; upper ones dried, lower ones still moist.

1¼"

½"

Woodrats, or packrats, as they are sometimes called, include 12 species within the genus *Neotoma,* and are pretty well distributed over western North America, from the southern border of Yukon Territory south into Mexico. They occur more sparingly in the eastern states. Woodrats live in a great variety of places, some species choosing a home among the cacti of the deserts, others preferring cliffs or rockslides, still others the forest floor.

The woodrat feet have fairly stubby toes with short claws, which may or may not show in tracks of animals walking or moving slowly. Tracks are very similar to those of white-footed mice, yet on a larger scale. The front feet have five toes, but the innermost toe is greatly reduced and appears as just a dot on the inside of tracks halfway between the palm and hindmost pads. The hind track may show up to six distinct pads when the heel fully registers, as when woodrats bound. The front tracks of white-throated woodrats, *N. albigula,* measure ⅜–⅝ inch long and ⁷⁄₁₆–⅝ inch wide. Hind tracks are ⁷⁄₁₆–¾ inch long and ⁷⁄₁₆–¹¹⁄₁₆ inch wide.

These animals are most easily discovered by their bulky nests, which may conceal a burrow entrance or harbor one or more bowl-shaped nests of shredded bark or grasses. The nest may be a pile of sticks and miscellaneous debris of moderate size in a bush, or back in a crevice of a cliff or cave, or a huge pile of sticks rising to a height of 4 or 5 feet or higher in a thicket or up against a tree.

(Left) *A dusky-footed woodrat walks along a riverbed in central California.* (Above) *The large nest of a dusky-footed woodrat in central California.*

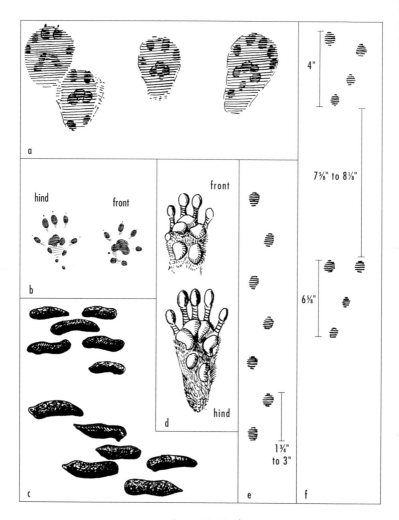

Figure 52. Woodrat

a. Tracks in mud of *Neotoma albigula,* natural size.

b. Front and hind tracks of *N. albigula* in wet sand (Texas).

c. Scats, natural size (upper, from Chisos Mountains, Texas; lower, Jackson Hole, Wyoming).

d. Feet of a woodrat from British Columbia, *N. cinerea,* natural size.

e. Walking track pattern.

f. Bounding pattern (Nevada).

Some are built in trees, up as high as 20 feet or more. In the desert a nest is often built into a clump of cactus or mesquite. These houses, whatever their location, usually include in the material a collection of odd objects—an old spoon, several bones, horse dung, tin can, almost anything that catches the fancy of this unusual architect. If there is a pile of rubbish in a corner of an abandoned cabin, the nest may lie snugly on top of the collection, in the open. In fact, some of these structures look more like a pile of junk hoarded by a collector of heterogeneous curios than a domicile. And some nests are the collection not only of one woodrat but of many, as nests can be used by successive generations of woodrats in an area. Woodrats also inhabit unadorned burrows in the ground, as is common with the bushy-tailed woodrat, *N. cinerea.*

The scats (figure 52, c) are fairly large, varying somewhat in size with the species. It has been reported that there is a slight size difference between the scats of the sexes. They are generally plentiful in the vicinity of the house or cliff abode, often accumulated in heaps. In many areas, especially on cliffs near the home crevice, there is an accumulation of dung that has a homogeneous, tarlike consistency, black in color—a mass that may be 6 inches or more in diameter. In some instances this black deposit drips over the edge of the rock and extends down the face as a somewhat sticky overflow for a distance of a foot or more (see figure 53). In the same areas are white spots or streaks formed by accumulating residue of urine. On little rock edges smeared with this whitewash type of stain, you will find little honey-colored

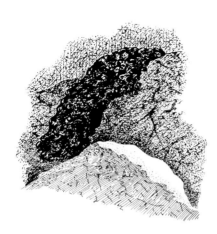

Figure 53.
Black tarlike material deposited on rock ledges by woodrats, apparently an accumulation of a soft type of droppings. The white stain on the rock edge is caused by accumulated calcareous deposits from repeated urinations.

drops, or streaks, along the ridge. So prominent are these stains that in some western states it is possible to see the white woodrat spottings on cliffs while driving by on the highway. They may be distinguished from the familiar whitewash stains left on rocks by ravens and some hawks and owls in this way: stains left by birds tend to be vertical streaks and heavily concentrated below a nest site or favorite perch; those left by woodrats tend to be shorter, and may be vertical, diagonal, or horizontal, thus presenting an irregular pattern on a cliff face.

Often, if you find a woodrat at home in his rubbish domicile, he will respond to a disturbance by drumming or thumping with the hind feet, both coming down together. You may hear this if you are in a house or cabin that has been invaded by a woodrat. Or he will drum by pounding his tail. This and thumping with the feet appear to be nervous reactions to disturbances, though they may have a more specific meaning in woodrat communications.

The woodrat personality—native American, antique collector, at home in the cactus desert and the northern forest as well as on the steepest cliffs and in rocky caves—is well worth cultivating. The traces he leaves are the means toward acquaintance.

NORWAY RAT

The Norway rat, *Rattus norvegicus,* is one of the exotic rodents that has made itself at home over the North American continent, chiefly at human habitations.

Apparently this common creature, so parasitic on human food supplies, has not been an attractive subject for study by naturalists. However, R. G. Pisano and Tracy I. Storer made such a study and recorded a de-

Bounding tracks of a Norway rat along a riverbank in western Massachusetts.

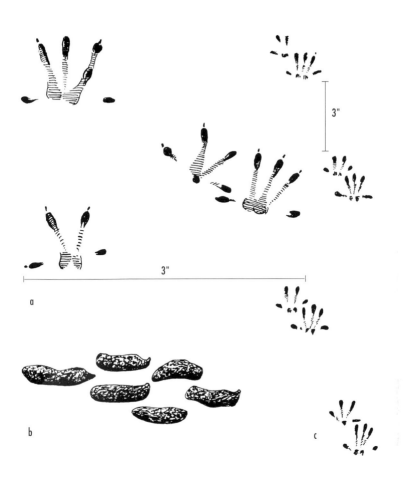

Figure 54. Tracks and scats of Norway rat

a. Tracks, in bounding pattern, on mud surface, about natural size. Distance between jumps 7½ in. The heels of the 5-toed hind feet did not show.

b. Droppings, natural size.

c. Walking track pattern. Note the 4-toed front and 5-toed hind feet.

scription of the burrows and other evidence of their presence. The burrows were 2–2½ inches in diameter. It was also found that burrow entrances would sometimes be plugged with earth, much as a pocket gopher would do it.

Like other rodents, rats will make runways through the vegetation. On Atka and Rat Islands in the Aleutian chain of Alaska, rats have adapted themselves to a natural habitat. There you will find their burrows and runways with little piles of cut grass segments in the manner of our native meadow voles. On Rat Island, where vegetation is rather scarce, the rats have found refuge in the boulder beaches. Some have had the audacity to dig their burrows into the foundations of the cliff nests of the bald eagle! Perhaps they are clever enough to pilfer food scraps from the nest itself. Tracks and scats are shown in figure 54.

HOUSE MOUSE

Mus musculus, an immigrant from the Old World, is a sophisticated creature that has learned to thrive in human homes, in the cupboards, the barns, the dumping ground, among the cages at the zoo—wherever people have left food or refuse. Although this mouse does at times venture into the woodlands, it is primarily commensal with humans.

Consequently, you will not commonly find the tracks in the woods. If you do, they will be quite similar to those of the white-footed mouse, as you see by the accompanying sketches. Note the similarity of patterns in mud of the slow run, or scurrying, shown in figure 55, b, for the house mouse and in figure 49, f, for the white-footed mouse. There is a difference in the position of the hind and front tracks in the two, but this could easily be a variation that occurs with either mouse. Note that in each case the hind track is recognized by the three forward-pointing toes and the two at the sides—five toes—while the front track has four, two forward and two lateral. Note also the similarity in the bounding gaits of the two in snow (figure 55, c and d, and figure 49, c and d).

It is interesting that the common house mouse, so much despised as a pest, occasionally gives voice to a birdlike trilling song. This has been described by observers from many parts of the world. However, the song is very rarely heard, and only at a distance of a few feet.

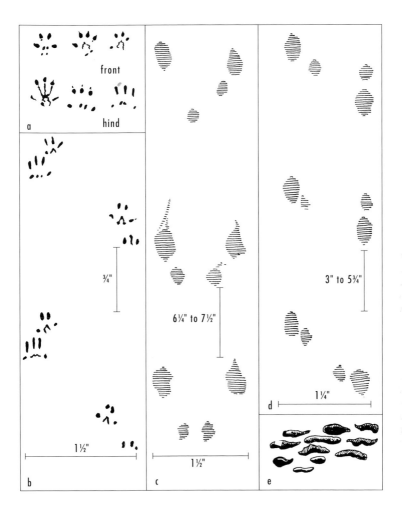

Figure 55. Tracks and scats of house mouse

a. Sample tracks in mud, natural size. With such prints the sole of the foot is scarcely indicated.

b. A "scurrying" gait — a modified bound — in mud, natural size; somewhat similar in pattern to the run in snow as shown in d.

c and d. Bounding gaits in snow; the more typical of the fast bound is that shown in c.

e. Droppings, about natural size.

There are 26 rodent species living north of Mexico that currently carry the name "vole." Add the 2 bog lemming species and that brings us to 28. How can one properly characterize a group of rodents made up of so many species, many of them specialized in habits, and varying in overall length from about 4 inches in the smaller forms to 7 inches? We can surely say that the meadow vole, *Microtus pennsylvanicus,* is more widespread over America, from Mexico to the Arctic, than any other rodent group. Also, in general appearance the numerous species and subspecies are similar—chunky build, long bodies, rounded noses, and ears more or less snuggled into the fur. Those represented here are suggestive of the group.

Meadow voles and white-footed mice together furnish most of the tracks one finds in winter snow. But in heavy snow country the meadow voles tend to remain beneath the surface, where they travel in snowy tunnels and have their snug nests.

In their typical patterns the tracks of the vole differ greatly from those of the white-footed mouse. The white-footed mouse ordinarily leaves a leaping four-print pattern (see figure 49, c), whereas the leaping vole generally leaves a double-print type of

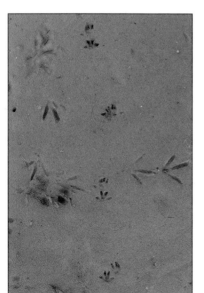

trail (figure 57, g), similar to that of weasels in diminutive form. But the vole leaves a variety of track patterns, depending on depth of snow, character of mud, sand, dust, gait, and speed. Look at some of the patterns in figures 56, 57, and 58.

The form of the footprint itself is shown in figure 56, a and b, in mud, and in figure 57, a, in the leaping pattern in

The trotting trail of a California vole, a western counterpart of the meadow vole, among western gray squirrel and spotted sandpiper tracks, in Sacramento Valley, California.

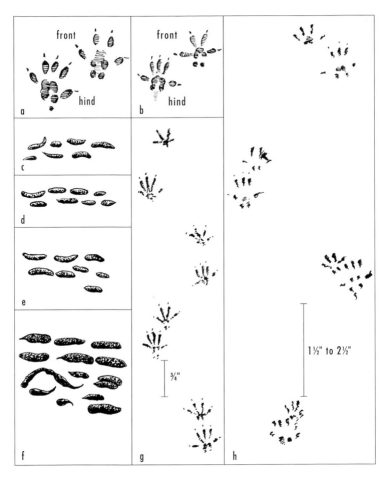

Figure 56. Meadow vole sign

a. Tracks in mud of the big water vole, *Microtus richardsoni*, natural size (Wyoming).
b. Tracks in mud of the montane vole, *M. montanus*, natural size (Wyoming).
c to f. Droppings, natural size: c, *M. oeconomus* (Alaska); d, *M. montanus* (Wyoming); e, *M. miurus* (Alaska); f, *M. richardsoni* (Wyoming).
g. Trail pattern of *M. montanus* walking in mud; straddle 1¼ in.
h. Trail pattern of *M. richardsoni* walking in mud; straddle 2 in.

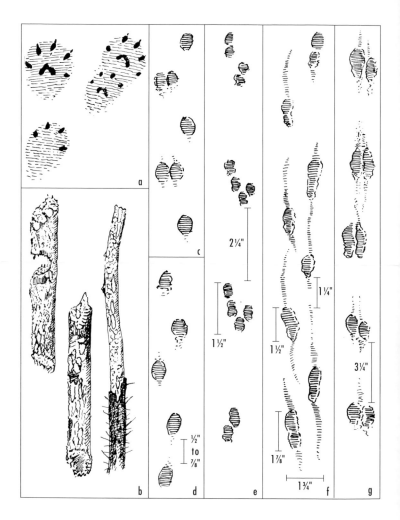

Figure 57. Tracks of montane vole

a. Tracks of *Microtus montanus* in light snow, natural size (Wyoming).

b. Wild rose twigs gnawed by *M. montanus,* natural size (Wyoming).

c and d. Tracks walking/trotting in light snow.

e. Loping in snow.

f. Trot, in snow.

g. Bounding in snow, a common track pattern.

light snow. Note the four toes of the front foot and five toes on the hind. Figure 57, c and d, illustrate the common track pattern of meadow voles: a trotting pattern in a thin film of snow, the hind foot registers in the front track, though in c sometimes one print is left beside the other. The same tendency is shown in f, where the animal was traveling in snow and the hind foot left a print a little behind the front track. Here the snow shows toe drag marks and front and hind tracks tend to connect. When we speak of a vole trotting we must not visualize the slow stepping of a raccoon or deer; voles share the same body mechanics as foxes and other canines when they are trotting and bouncing along. But when they leap, or bound, they leave a track pattern such as in e or g, in which the bounds may measure 2–6 inches ordinarily and in the case of a large individual, as much as 19 inches at extreme speed. These tracks are those of an average-sized montane vole, *M. montanus,* in Wyoming. The tracks of the large water vole, *M. richardsoni,* the adults of which suggest a young muskrat in size, are shown in figure 56, a and h. For meadow voles, front tracks measure ¼–⁷⁄₁₆ inch long and ⁹⁄₃₂–⁷⁄₁₆ inch wide. Hind tracks are ³⁄₈–⅝ inch long and ⅜–⁹⁄₁₆ inch wide.

In most cases the droppings of voles are distinctive among rodents, but they vary somewhat with diet. Vole scats are tubular, very smooth and with rounded ends, as compared to the twisty, tapered scats of mice. Scats of several species are shown in figure 56, c through f. Those of the big *M. richardsoni* are enough larger so that they may be fairly well identified in the field.

An important feature in the life of the meadow vole is the runway in the grass. Part the heavy grass cover where these mice are living and you will find these little pathways cut through the grass, radiating from a burrow, leading from one bush to another, or otherwise furnishing the traffic lanes to where the little grasseaters want to go. In these runways, too, you will find little piles of cut grass stems where they have been feeding—a good sign of the vole. In the winter the vole will gnaw the bark of bushes. Figure 57, b, shows the gnawed twig of wild rose. The tooth marks are much finer than those of the rabbit, beaver, and other gnawing animals, but, as shown in the rose twigs illustrated here, very often the bark or wood is chipped out without leaving clean-cut tooth marks. One must judge the work by the generally fine-marked appearance.

In snow country the vole lives below the snow surface, coming out only occasionally, and for such a life it builds a nest on the surface of the ground. When the snow thaws in spring you will find these globular surface nests constructed of dry grass, with an entrance at one side.

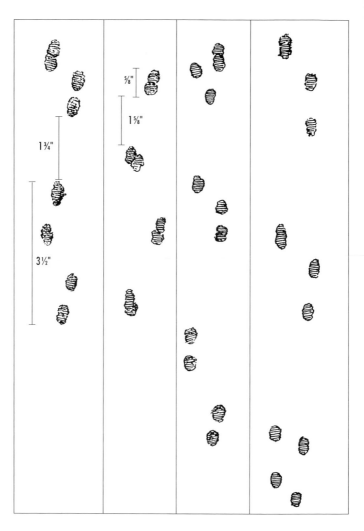

Figure 58.
Some sample segments of the trail of a meadow vole that had scrambled across the frosty surface of hard snow. These samples were taken from a section of the trail about 20 ft. in length. In this short distance in the travel of the same individual meadow vole were found all possible variations, from the 2-by-2 pattern to the 4-print pattern shown at the end. The outside width, or straddle, was 1 1/16–1 3/8 in. The jumps were 1 3/4–4 1/4 in. (March 5, 1953).

Sometimes too you will find ropelike cores of dirt and grass cuttings mixed together which are the material that had been shoved into snow tunnels in winter (figure 59), similar to dirt cores left by pocket gophers.

In spring voles move to underground nests in a subterranean burrow network. Some species will build their globular summer nests in tangled stems of reeds or cattails over the surface of water, and swim readily to and from the nest. If you are wandering over a flower-studded meadow in some of the western mountains, look closely along some of the small winding meadow streams. You may see a sizable mouse dash along the edge of the water into

Figure 59.

Cores of dead grass and other debris that meadow voles had pushed into snow tunnels in winter. This material is scraped out of shallow grooves in the ground, and small pits where the voles have been digging. Sometimes these cores consist of grass, sticks, and dirt; they are from 1 to 3½ in. in diameter. Often in spring the thawing snow will reveal these runways and cores radiating from an abandoned winter nest. Sometimes the cores of the meadow vole will be intermingled with the larger ones of the pocket gopher.

a bank burrow, or even into the water. This is the home of the water vole, *M. richardsoni*, mentioned previously. Some entrances to the burrow are under water, for these mice are quite aquatic in their habits.

Then again you may be traveling across a sage flat in the Cascade Mountain region, or on the rocky scree above timberline. Here you may find the network of tunnels and burrow entrances of the western heather vole, *Phenacomys intermedius*. Sometimes you will find their latrines and feeding signs under rocks.

In some of the arid sagelands of the Southwest you may find a green oasis where a small spring oozes out. Look about in the green grass and you may find vole runways there.

A number of species are known to store up food for winter, especially roots of plants. In North Dakota this habit was so notice-

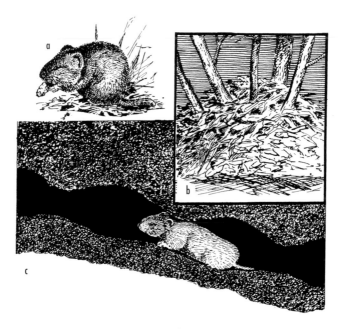

Figure 60. Singing vole, *Microtus miurus*

a. Singing vole, or hay mouse, munching a root.
b. A haystack in the base of a willow, stored by the singing vole.
c. Underground burrow, showing the narrow passages between the larger portions.

able that a variety of meadow vole, *M. pennsylvanicus,* became known as the bean mouse because of its underground storage of the beans of *Falcata comosa* and the tubers of a wild sunflower. My brother found a store of 706 tubers of bistort gathered by a heather vole and a cache of dandelion and other roots stored by *M. montanus.*

Not long ago, when we were traveling over the open tundra slopes of Denali National Park in Alaska, my brother showed me the workings of one of the most remarkable voles, described as *M. miurus,* whose life history he had been studying. We examined elaborate root caches stored away under the moss. There were, in addition, heaped-up stores of dried plants, some of them nearly a bushel in volume. So extensive is this haymaking that we have been calling this singing vole the "hay mouse" ever since. Dried material included grass as well as the stems and leaves of many other plants. As we fought our way through willow patches, we found many willow tips nipped off, and found the dried leafy twigs in the miniature haystacks nearby. The voles had climbed up among the willow stems to harvest the twigs. These haystacks were similar to those of another haymaker, the pika, but the pika lives up in the rockslides. The singing vole piles his around the base of a willow or the roots of a spruce tree (see figure 60, b). Furthermore, if you excavate its burrow, you may find that it consists of a series of larger sections connected by narrower passages.

This can be but a brief introduction to the myriad rodent populations of the continent that we know as voles, or meadow mice.

WOODLAND VOLE, SAGEBRUSH VOLE, BOG LEMMINGS, AND RED-BACKED VOLES

These members of the vole family are similar in form to the common meadow vole.

The woodland, or pine, vole, *Microtus pinetorum,* is more common in deciduous woods than pine woods. In Oklahoma I found them in oak woods and open grasslands. This animal is really a meadow vole somewhat specialized for subterranean existence,

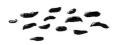

about natural size   greatly enlarged

Figure 61. Droppings of red-backed vole

for it lives pretty much in underground burrows a few inches below the surface. It makes underground storage of roots and tubers, and is very prone to gnaw the roots and bark of small trees and shrubs. Examination of plaster casts of their tracks reveals no practical difference from the individual tracks of *Microtus* shown in figure 56, g.

The sagebrush vole, *Lemmiscus curtatus*, inhabits the western dry lands. They have runways and burrows, and leave cut grass stems just as do the common meadow voles. Their tracks are similar to meadow voles; I've only seen clear tracks on two occasions. The claws seemed more pronounced and the toe pads smaller than those in tracks of meadow voles.

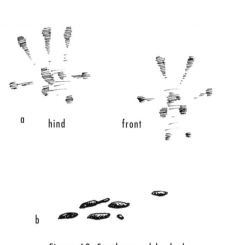

a   hind   front

2"

b

Figure 62. Southern red-backed
vole tracks and droppings

a. Tracks of *Clethrionomys gapperi* in mud (New Hampshire).

b. Droppings are tiny, resembling those of common house mouse. They vary with diet and size of animal. Natural size (drawn by Carroll B. Colby from specimens in the American Museum of Natural History, New York).

c. Trail in light snow, vole changing from a paired loping gait to a trot (New Hampshire).

c

The bog lemmings, *Synaptomys sp.*, are short-tailed voles of the northern and eastern part of our continent. I have not found their tracks so as to be sure of their identification, but they could not differ materially from those of the common meadow vole. They are known to use the same runways with *Microtus* at times. Their droppings are rather small, but not distinctive.

The red-backed voles, *Clethrionomys sp.*, are also members of the vole tribe which share the general form of *Microtus*. They are common in the northern forests of the continent, as well as in the western and eastern mountains farther south. You are likely to find them among stumps and logs and forest litter, or in the mossy muskeg country of the North. I have not found this vole building elaborate runways in the vegetation to the extent that the meadow vole does. It runs about more freely, like the white-footed mouse. Its tracks and trails are similar to meadow vole patterns, but smaller; the droppings (figure 61) are smaller than those of *Microtus* but otherwise not distinctive. Figure 62 shows tracks and trails of the southern red-backed vole.

TREE VOLES

One day many years ago, in the heavy woods near Forest Grove, Oregon, I climbed a fir tree to look at a nest some 20 feet from the ground. It was an old nest, possibly built by a hawk. But I was puzzled to find it filled completely by shredded fir needles. It was not until a year later that I learned that I had obviously come on the nest of the red tree vole, or tree phenacomys, *Arborimus long-icaudus*, an animal restricted in range to the humid forests of coastal Oregon and northern California. The everyday, common meadow vole, *Microtus pennsylvanicus*, which we have discussed already, lives on and in the ground. Here is a mouse of a related genus of voles that has followed the example of the harvest mouse, but that has gone much farther and builds a nest high in the trees. Not only that, it has specialized its diet chiefly to needles of fir, spruce, and hemlock and eats only the central strip of the needle, leaving the two edges for building a nest. Like the porcupine, and probably other animals, it prefers the young terminal needles, and eats the whole needle in this case. Nests may be original structures built by the tree voles themselves, close to the trunk of small trees or out on big heavy limbs of large trees, or this creature might build its structure into an old bird's nest, abandoned squirrel nest, or woodrat nest. It is roughly globular, with entrance in the side. It varies in size from a few inches in diameter to a large mass 2 feet or more in diameter with several entrances. The nests built from scratch by the mice are made on a ground structure of fine twigs and possibly lichens, filled out with split conifer needles. They may be placed from 10 to nearly 100 feet above the ground.

COMMON MUSKRAT

The common muskrat, *Ondatra zibethicus,* is so widespread over the continent that only the more desertlike regions and the true Arctic are outside its range.

Muskrat tracks are, of course, found mostly near water, though occasionally the animals will make fairly long overland journeys. Both feet have five toes, though the inner toe of the front foot is so small that it rarely shows more than a claw. See the minute imprint in figure 63, a. Look for lateral lines running across the hind tracks, formed by the unusual pads on the feet, as well as a ridge created by stiff guard hairs surrounding each toe. Front tracks measure ⅞–1½ inches long and 1–1½ inches wide. Hind tracks are 1½–2¾ inches long and 1½–2½ inches wide.

In the track pattern the front foot very often precedes the hind foot. At other times the hind foot is foremost, or may cover the front track. Examples are shown in figure 63. Very often, too, the drag mark of the tail is shown, but not always. The width of the walking trail, or straddle, is between 3 and 5 inches.

One winter day at a sluggish stream I watched a muskrat running hurriedly across a sheet of ice covered with a little snow. The tail moved about, sometimes swinging high aloft. The trail in this hasty gait was very irregular, with only an occasional tail mark, and with front and hind feet in many arrangements, as shown in figure 63, e.

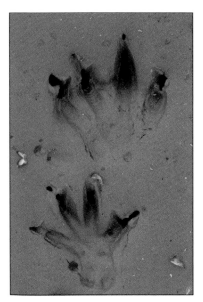

The smaller front and larger hind tracks of a common muskrat along the Sacramento River in California.

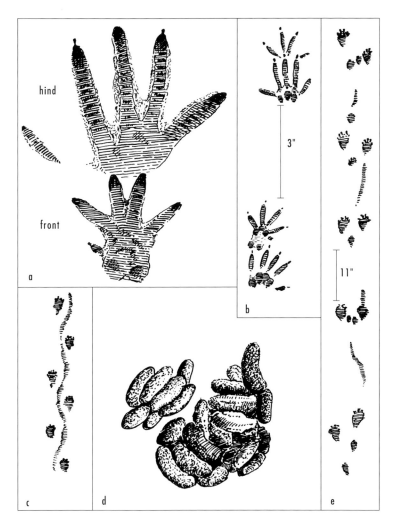

Figure 63. Muskrat tracks and scats

a. Tracks, right side, in mud; natural size.
b. Walking tracks in mud.
c. Walking trail, showing tail mark in snow.
d. Droppings, about natural size.
e. Bounding trail in light snow on ice.

Muskrat droppings are elongated, varying from ⅜–1 inch in length. They may be found in clusters on logs in the water, beaver dams, rocks, or favorite resting places on the bank (see figure 63, d).

There are other signs of muskrat. Prominent among these is the muskrat house, built of matted vegetation, on the shore at the water's edge or in shallow water. It consists of marsh grasses and sedges, sometimes heaped up to as much as 4 feet in height. Then there is the food shelter, or eating house, which may be built over a "plunge hole" in a marsh, or on the ice in winter. In areas where the winter is cold enough, the muskrat will push up through a hole in the ice various types of debris, forming a mass in which a cavity large enough for one animal is formed. This becomes covered with snow, with enough insulation to keep the plunge hole in the ice open. Figure 64 shows such a hut on the ice, together with a bank burrow, all in cross section.

In waters used by muskrats you will see floating blades of sedges or other food remnants left by these animals, though this is also true of nutria. It may be a collection of cattail stalks, apparently cut into convenient lengths for handling, much as you may find lengths of grass stems in field mouse runways. Or there may be a floating raft of cut stems of various kinds, on which the muskrat rests to feed. In dense masses of marsh vegetation, too, you will find holes where muskrats have dug for roots. They often go ashore on favorite spots to feed, where you will find fragments of stems and blades.

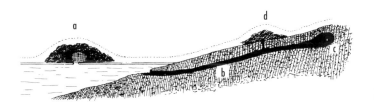

Figure 64. Cross section of muskrat burrow and
feeding platform, all under a layer of snow

a. Feeding station on the ice, with chamber and plunge hole down through the ice.

b. Simple burrow, with underwater entrance and nest chamber, c, at the end above water level.

d. Pile of vegetation covering an accidental hole (?) or ventilation hole that leads down to the burrow.

The muskrat is not entirely vegetarian, and you may find heaps of clamshells on a bank or in shallow water, on favorite feeding spots, where the animals have opened the mollusks and fed on them.

The muskrat often builds bank burrows, with underwater entrance. The home burrow may be complicated, with various passages and a nest chamber. There is also a simpler type used for a refuge or retreat—in fact, it may be simply a shallow cavity under a bank where the animal can rest and feed. In some instances I have found an opening covered with vegetation leading up from the tunnel to the open air on the bank. Was this an accidental hole that had to be covered over, or was it intentional? The diagram in figure 64 illustrates such a one found on the Koyukuk River, Alaska, in 1924.

There is still another sign that should be mentioned. On the bank near the water you may come upon a little platform or mat of cut sedge leaves or similar fragments, sometimes mixed with a little mud. This is a scent post, for the muskrat also has scent glands and leaves an odor record for other muskrat passersby. I have never found them large or elaborate, such as those made by beavers. It may be simply a handful of stems and a little mud. Occasionally I have found a little mat of stems without mud, which may or may not have been a scent post. Their scats may also adorn such structures.

The muskrat voice seems to be mostly a squeaking, which I have rarely heard.

It is worth mentioning, too, that although muskrats are especially partial to swamps and marshes and certain streams and lake shores, they are prone to seek the beaver ponds as ready-made habitat for their homes. They appear to associate regularly with beavers in this way. Muskrats will also live in saltwater bays and even on nearby coastal islands.

COLLARED AND BROWN LEMMINGS

One winter evening near the Brooks Range in Alaska, my brother and I were talking with a group of Eskimos. They told us about Kilyungmituk, "the little one who came down from the sky." It seems that there was a bear up in the land of the sky, but it began falling, and the closer to earth it fell, the smaller it became, until it plopped into the snow in the form of the lemming. "We know this," Pooto declared, "because his tracks are like small bear tracks." Selawik Sam spoke up and gravely told us that he, himself, had seen the holes in the snow where the lemmings had landed.

The trail of a collared lemming in Churchill, Manitoba.

Here, then, is an instance where a rodent track has woven itself into human legends. In figure 65, d, are shown the front feet of a collared lemming, *Dicrostonyx* sp., with the peculiar claws that no doubt produce an interesting footprint if the details show. I can only show the trail patterns, however, adapted from photographs kindly lent by Charles O. Handley Jr., who obtained them in the Canadian Arctic, and from photographs and drawings by Dr. Robert Rausch, of Alaska. It will be noticed that these lemming trails (figures 65 and 66) are very similar to those of the common meadow vole, as shown in figures 57 and 58.

Finding lemmings is fairly easy where they are plentiful, for they have their burrows in the mossy tundra, but I had a particularly frustrating experience on Unalaska Island in the Aleutians, where specimens of the little-known lemming of that island were particularly desired for scientific studies. I searched diligently and trapped in likely places for several days without getting a single specimen. During the same period I was also collecting red fox scats for analysis, to determine the fox diet. It was exasperating to find lemming remains in many of the fox scats. Evidently a keen fox nose and training in nature from puppyhood can outdo the efforts of the field biologist.

The collared lemming, found across the Arctic, turns white in winter. However, the species that have developed in the southward extension along the Bering Sea in the Aleutian chain do not turn white.

Lemmings are some of the most interesting animals of the Far North. Periodically hordes of lemmings in Europe move over the lowlands toward the sea, into which they finally plunge and drown. The journeys seem to occur when the lemming population has become too great for its natural habitat.

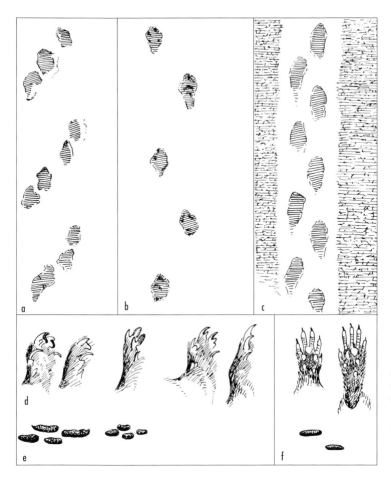

Figure 65. Collared lemming and brown lemming

a. Loping track patterns of the collared lemming, *Dicrostonyx* sp., in snow, from the Canadian Arctic.

b. Trotting. In c the tracks are in a snowy trough formed by the body.

d. The front feet of collared lemming: at left, showing the "double" middle claws developed in winter; at right, the summer claws, from which the underpart has been shed.

e. Scats of *D. groenlandicus*, showing variations; natural size.

f. Feet of the brown lemming, *Lemmus trimucronatus*, which do not produce the enlarged claws in winter. Below are typical scats, natural size.

The brown lemming, *Lemmus trimucronatus*, is found in the tundra and the boreal forests. It does not change to a white coat in winter as the collared lemming does, nor does it grow the peculiar "double" claws in winter. You will find it pretty much in the same habitat as that of some northern voles of the genus *Microtus*.

In northern Alaska I was told of certain mass migrations of "reddish mice," similar to the migrations of the collared lem-

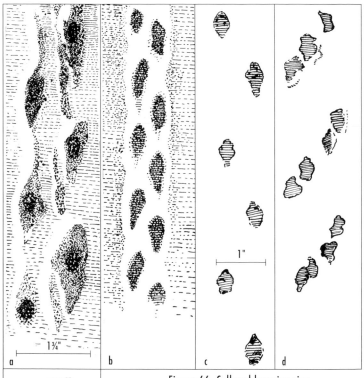

Figure 66. Collared lemming sign

a. Trail in snow, from Alaska. Individual footprints, ⅜ in. wide by about ½ in. long. Width of the trail pattern, 1¾ in.

b. A similar trail in snow (Canadian Arctic).

c. Trotting trail in snow with 1⅜ in. between footprints; straddle slightly over 1 in. (Alaska).

d. A loping gait.

e. Scats.

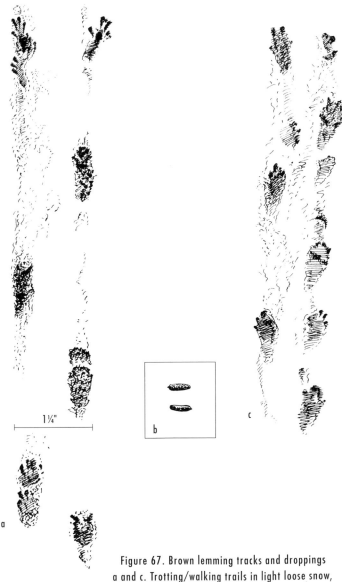

1¼"

Figure 67. Brown lemming tracks and droppings
a and c. Trotting/walking trails in light loose snow,
    showing irregular pattern, about ⅔ natural size.
b. Droppings, about natural size.

mings. These reports may have referred to the brown lemming. In the Yukon delta region I have seen the brown lemmings in spring among the stranded ice blocks along the coast, but did not see them actually take to the water.

In figure 67 are shown some brown lemming tracks in light snow, with rather irregular pattern, suggesting similar patterns of the meadow vole.

JUMPING MICE

Jumping mice are not well known to the public, yet they are distributed over most of the United States—northward into Canada and Alaska, in some regions as far as the Arctic Circle; southward in the West into California and Nevada; and in the East to North Carolina. There are two genera: the more widely distributed *Zapus*, with three forms, and the larger *Napaeozapus*, confined to eastern Canada and the eastern states.

Jumping mice are frequently found in meadows and grassy mountain slopes, though they live in woodlands as well. Generally you are aware of their presence by seeing a small creature leap away from you with half a dozen froglike jumps, then suddenly stop and apparently vanish. Many times in such instances I have crept up carefully to where I last saw it and spied it crouching there in the thick grass or border of a bush. Then, when in range, I have pounced on it with cupped hand to look it over before releasing it.

In Ungava I have seen the large *Napaeozapus* leap into the water and dive when frightened. It appears to frequent streambanks quite commonly.

These mice make long leaps, some recorded to be from 6 to 12 feet, or more. I have never seen them leap more than an estimated 4 or 5 feet; sometimes they only make a hop of 2 feet or less.

You will not find the tracks of this mouse in the winter snows, for they spend the winter in hibernation. In summer their tracks are common in puddles and along banks of water sources where

Figure 68 (opposite)
a. Tracks in mud, natural size. It is unusual that the heels have so well registered.
b. Right front and hind feet, underside; slightly over natural size.
c. Typical tracks in mud, where the hind heel did not touch the ground; natural size.
d. Droppings, natural size. These are all from the same individual.
e. Tracks in mud, on slow hopping gait, tail drag continuous; about ⅔ natural size. The registration of both heels and tail is very uncommon.

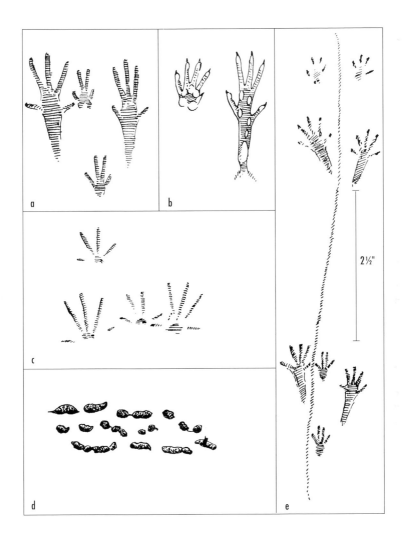

Figure 68. Tracks and scats of jumping mouse

Perfect tracks of a meadow jumping mouse along the Connecticut River in Massachusetts.

substrates are soft enough for light creatures to register well. A pet jumping mouse left the tracks in mud which are shown in figure 68. Clear tracks will show the "ribbed" digits, similar to those found in meadow vole and larger muskrat tracks. Front tracks measure ⁷⁄₁₆–⅝ inch long and ⅜–⅝ inch wide. Hind tracks are ⁷⁄₁₆–⅞ inch long and ⅜–¹¹⁄₁₆ inch wide.

Although the jumping mice do not make runways in the grass, you will find the little piles of grass stems where they have been feeding, similar to those left by meadow mice but usually longer. Also, they build a globular nest on the ground, or at the end of a short burrow.

NORTH AMERICAN PORCUPINE

The North American porcupine is found in the coniferous forests of Canada and Alaska, and down through the timbered areas of the West, as far east as the western part of the Dakotas and northwestern Texas. It is also in the north woods of the eastern states, around the Great Lakes, and as far south as West Virginia.

Officially known as *Erethizon dorsatum*, the porcupine plays the role among rodents that the skunk has among weasels. Both have developed such a strong defense that they require neither speed nor agility. The skunk when disturbed can spray a vile-

Figure 69 (opposite)
a. Footprints in mud, a little less than natural size. Note that the front feet have 4 toes and the hind 5.

b. Trail in snow, showing drag marks of feet.

c. Trail in dust, showing brush marks of quills.

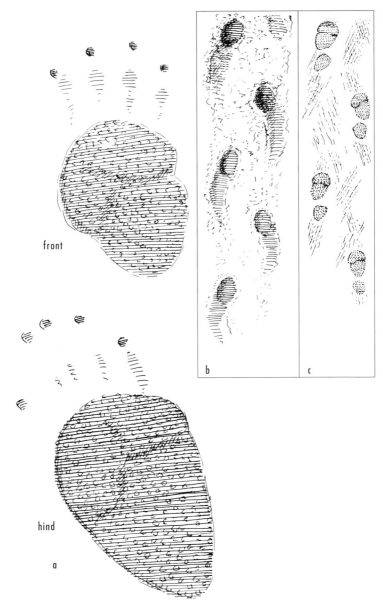

Figure 69. Porcupine tracks

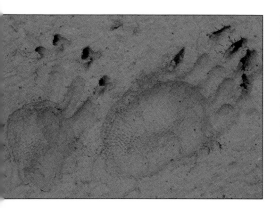

Exceptional front and hind porcupine tracks in south-central New Hampshire, yet somewhat unusual in that the toes have registered so clearly; often only the claws are evident.

smelling liquid. The porcupine can curl up and bristle like a live pincushion. It need only find some secure place in which to shove its nose, for the remainder of its body is well protected. When it is touched or approached closely, the exceedingly sharp quills are raised and readily come loose from the skin. The quills cannot be thrown, as some people imagine.

The waddling gait of this animal reveals a great deal. One day a group and I were walking up a well-worn trail in the Teton Mountains of Wyoming. In the dust we saw at frequent intervals little brush marks, as if someone had gone that way with a whiskbroom. It was clear that a porcupine had passed by. The heavy stiff tail swings back and forth with the clumsy gait, and the wiry hair on the underside lightly brushes the ground, producing those zigzag whiskbroom marks in the dust. They are not always prominent, and sometimes, on ground where the footprints themselves do not show, they are very faint. See figure 69. At other times the tail is held aloft and leaves no mark at all.

When you find the tracks in wet snow or mud, or heavy dust, several things are revealed. The front track has only four toes, while the hind has five. You will also see short toed-in steps, somewhat in the style of the badger. These prints are 5–10 inches apart. Also, there are marks far ahead of the print, denoting the long claws. Finally, note the pebbled appearance of the print itself, the "fingerprint" pattern of the heavy calloused soles of the feet (figure 69, a and c). Front tracks measure 2¼–3⅜ inches long (with claws) and 1¼–1⅞ inches wide. Hind tracks are 2¾–4 inches long and 1¼–2 inches wide.

In snow there are drag marks of the feet, sometimes connecting one print with the next, and in deep snow a wallowed trough is formed. Usually I have found the hind track in front of the front

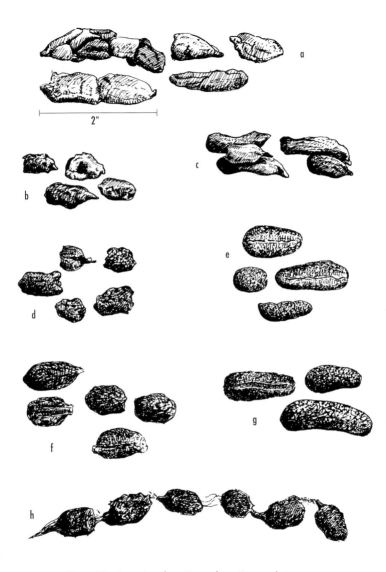

Figure 70. Porcupine droppings, about ⅔ natural size

A and c are the soft summer droppings (Wyoming). The others are various types of fall and winter scats: b and e from Jackson Hole, Wyoming; d from North Dakota; f and g from Alaska; h, from Wyoming, illustrates how pellets are sometimes connected when the food is semisoft.

track, although sometimes, as in figure 69, a, the front print is ahead. In snow the hind track usually registers in the front one. The width of trail pattern, or straddle, is 5–9 inches.

Porcupine droppings are variable, and looked at carelessly could be mistaken for those of deer. But note the characteristics shown in figure 70. The winter scats, of the pellet type, tend to be rough on the surface, and somewhat irregular. Occasionally they are perfectly smooth, as in e and g. A series is sometimes connected with strands of vegetation (h), as in the case of the marmot and some other rodents. The longer types tend to be curved slightly, and often on the inner curve there is a slight groove (e and g), which also appears on some of the rounder types (f). In summer, with succulent feed, the pellets become elongated and soft and tend to cling together in a mass (a and c). In b we have an intermediate stage—the short pellet, rather soft and irregular in shape.

During the winter porcupines tend to spend considerable time in a single feed tree and often one finds a large number of scats at the base. They are also found at and in caves in the rocks where the animals find refuge.

One evening when making camp I found a great quantity of pine twigs around the base of a tree, enough to make a comfortable bed on which to spread my sleeping bag. Around the tree were many porcupine droppings, and some barked areas on the trunk. The porcupine often nips off twigs, which can be found strewn on the ground below. However, be careful not to confuse these with twigs dropped by squirrels. The latter nip off cone-bearing twigs and then pick off the cones on the ground. The porcupine scats on the ground will tell the story.

The twigs thus dropped by porcupines are often wastefully abandoned after the animal has eaten only a portion of the foliage. Deer and elk often find these and other fallen limbs and accept them as food.

In winter porcupines eat the bark of twigs and gnaw large patches of bark off tree trunks. These have neatly gnawed edges, irregular in shape. Often a tree will be spotted with these gnawed patches, all the way up the trunk and on many of the limbs. They are distinguished from elk or moose barking by the neatly gnawed edges, irregular outline, and numerous small tooth marks; and of course if high from the ground, there can be no doubt who has been at work.

Porcupines are generally thought of as living in the forest, but they often occupy treeless areas, such as the Alaska Peninsula, where they find willows and alder brush for winter forage; and in summer they may be found out on the open high mesa, in mountain country—where they find refuge in rock crevices. I have found porcupines occupying caves in rocks, when these are available. In such places, at least those in the West, you may find a mixed accumulation of dung of the woodrat, marmot, and porcupine. They also seek cavities in the ground. I have found them using for shelter the dry portions of caved-in beaver burrows, near streams. And they may use a hollow tree.

These rodents will feed on the buds and catkins of willows and maples, the leaf shoots of aspens, and the nuts of beech and oak. In each of these cases porcupines nip branches in the tree, feed, and then drop the branch which they have freed. Cut branches will litter an area where they have been feeding and will be caught in the canopy of the tree in which they are sitting as well as in neighboring vegetation.

The porcupine generally appears silent, but can actually be quite vocal. One time I thought I heard the bleating of a moose calf, traced the sound to its source, and found two porcupines, one of them up in a hawthorn bush. Elsewhere I have described the calls in Alaska as "a combination snort and bark, an unexpected and startling sound heard on two occasions when the animal was approached suddenly; and a moaning sound, heard once, which my companion at first attributed to a cub bear." Others have described "screams" and crying like a baby's. At any rate, you may expect to trace such sounds to the porcupine.

AGOUTIS

Agoutis, of the genus *Dasyprocta,* are among the large rodents of South America that have pushed through Central America into Mexico. They have a humpbacked appearance, are tailless, and in attitude remind one of a large rabbit (without the big ears).

Once from a jungle trail on Barro Colorado Island in the Canal Zone, I saw a Central American agouti, *D. punctata,* running off across the forest floor into a thicket—the first I had seen in its native home. On the muddy banks of a small stream I found tracks where one had crossed. I saw the little three-toed prints in the muddy trail after a rain. Only in similar locations are you likely to find the footprints, for the jungle floor elsewhere is so littered with debris that such small animals are not apt to leave a clear trail there.

Strictly speaking, the front feet have five toes, the two outside ones very small, but the tracks you find commonly will appear three-toed (see figure 71).

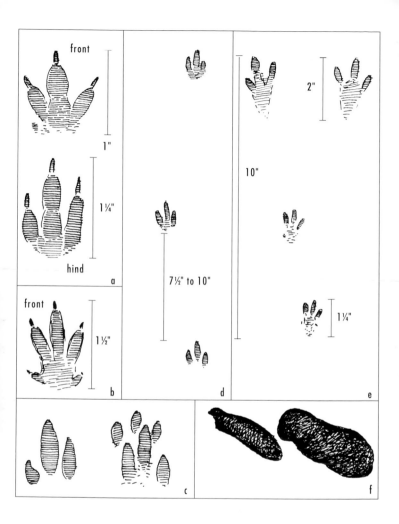

Figure 71. Agouti tracks and scats

a, b, c. Tracks in mud (Panama Canal Zone, 1952); c shows the fragmentary toe marks that one often finds.

d. Walking pattern.

e. Bounding pattern.

f. Scats, about ⅔ natural size.

PACA

The paca, *Agouti paca,* is one of those stodgy tropical rodents, gaily adorned with stripes and spots, that seem so incongruous with our notions of a rodent when we see them on display in the zoo. They live in the heavy vegetation of the jungle of tropical America, coming as far north as parts of Mexico.

When seen individually the front track shows four toes, the hind track three, but when the hind track is superimposed on the front one, as you will generally find them, you get an impression of a long-toed, three- or four-toed track (see figure 72).

The scats are in the form of consolidated pellets, as in figure 72, e, and sometimes in pellet form.

Front and hind paca tracks; photograph by Marcelo Aranda.

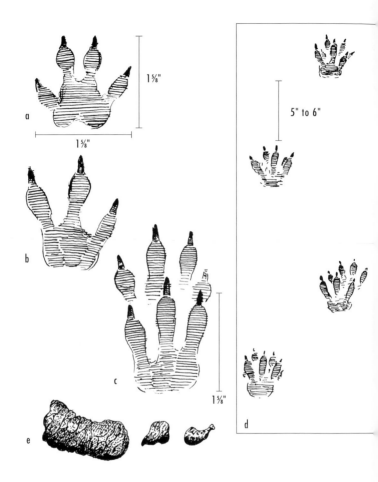

Figure 72. Paca sign, about ⅔ natural size,
the tracks in soft mud (Canal Zone, September 12, 1952)

a and b. Front and hind track, respectively.

c. Hind track partly covering front.

d. Walking track pattern.

e. Droppings.

FAMILY CANIDAE:
DOGS, FOXES, AND WOLVES

RELATED TO THE domestic dogs in America are the wolves, coyotes, and foxes. Their tracks have a similar pattern, with four toes, the toenails usually showing in the track (see figure 73). The front foot is slightly larger than the hind foot, and has five toes rather than four; the fifth is higher up on the inside of the leg and only registers at high speeds or in deep substrates. The heel pads of both feet, when shown clearly, have a fairly distinctive outline, the front one differing from the hind one, with some differences in shape among the members of this family. Note the more pronounced three-lobed rear margin of the hind pad, as compared with the front pad, in both the dog and wolf. In the coyote the outlines are different. The red fox has a curved ridge of callus across the pad, showing through the hair, and this may appear in a clear track, though generally the full outline of the hair-covered pad appears. In the gray fox the pads have a more pronounced hooklike projection on each side. Generally the Arctic fox does not show pronounced pad peculiarities, nor have I found them in the kit fox tracks.

The droppings of the dog tribe are remarkably similar. There are average differences in size, but the variation is so great that there is much overlapping in measurements and one cannot be certain of identification in some cases. Aside from variations due to types of food and quantities eaten, there is the confusing fact of smaller sizes created by young animals in each species.

The samples illustrated in figure 74 are offered as average or typical, for a guide to intelligent guesses in some instances, or more positive identification where it is known what animals are in the area and when one has gained familiarity with the animals' characteristics.

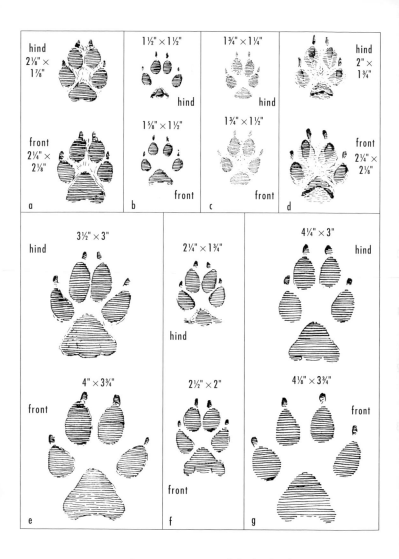

Figure 73. Footprints of the dog family

a. Arctic fox, in sand.
b. Common gray fox, in mud.
c. Kit fox, in snow.
d. Red fox, in mud.

e. Alaskan malamute, in mud.
f. Coyote, in mud.
g. Gray wolf, in mud.

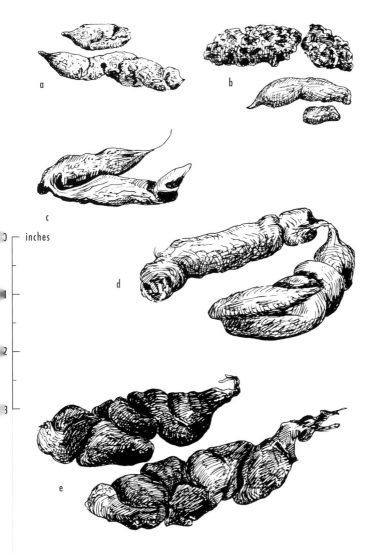

inches

Figure 74. Scats of the dog family
a. Common gray fox or kit fox (Texas).
b. Arctic fox (Aleutians).
c. Red fox (Alaska).
d. Coyote (Wyoming).
e. Gray wolf (Denali National Park, Alaska).

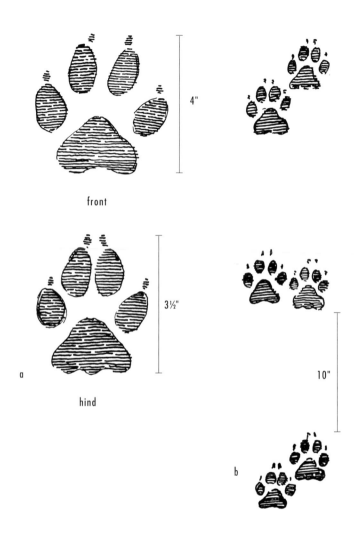

front

hind

a

b

4"

3½"

10"

Figure 75. Tracks of Alaskan malamute

a. Front and hind tracks.

b. The track pattern of a slow walk or pace, in which the front and hind legs on a given side of the body move nearly simultaneously.

Dogs are of such great variety of size and shape that it would be hopeless to characterize the tracks of all of them. In figure 75 are shown the tracks of the Alaskan malamute (a Husky dog) from Port Moller, Alaska. These could be confused with the gray wolf tracks, since both are found in the same territory. Figure 76 shows tracks of a cocker spaniel and a mongrel dog, for comparison.

There are certain features in the tracks of the dog family, common to dog, wolf, and fox, which might be stressed here. The heel pads of front and hind feet have distinctive outlines, but as you find the tracks in snow or mud there is a difference somewhat unrelated to the actual shape of the pad, a difference that helps to distinguish front and hind tracks: front and hind feet are held at

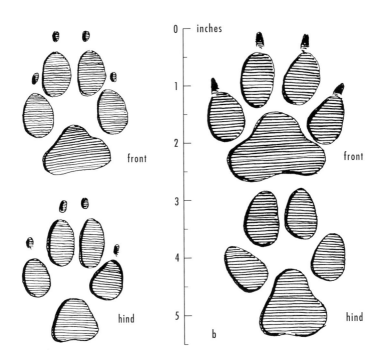

Figure 76. Cocker spaniel and mongrel dog tracks
a. Cocker spaniel, about natural size.
b. Mongrel dog, about ⅗ natural size.

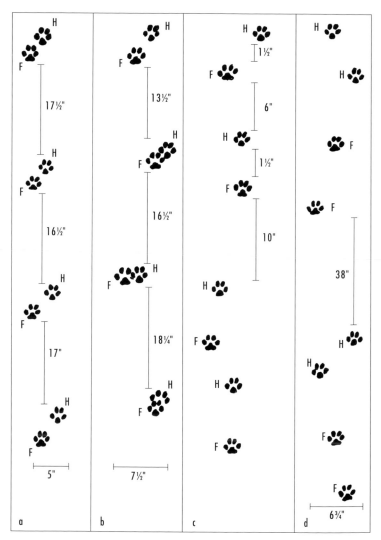

Figure 77. Mongrel dog trails in mud
a. Side trot, which is a faster trot; 5-in. straddle.
b. Another trot, with hind tracks nearly registered in front tracks.
c. Slow lope.
d. Gallop.

somewhat different angles as they strike the ground. Perhaps it is more important that the front paws in this group of animals are more mobile than the hind ones; and their toes tend to spread more. At any rate, the front heel pads make a fairly clear imprint, while the heel pad of the hind foot does not completely register in mud or snow, leaving out, or but faintly showing, its lateral lobes. Thus the hind-track heel pad tends to be merely a roundish or oval depression.

In figure 76 this is illustrated in both the cocker spaniel and mongrel dog tracks. Hind and front tracks can be distinguished also in the dog trails in figure 77. The same features can be found in coyote and fox tracks.

Note how in figure 77 the hind foot commonly oversteps the front track, though in b it tends to step in the front track at times. These gaits show a transition from a slow trot to the gallop in d, with c being an intermediate stage from a to d.

Compared to a coyote, a dog of like size would tend to have a shorter stride and larger tracks.

Dog droppings are very similar to those of the wolf, coyote, or fox, varying in size with the breed of dog and character of food. For distinguishing scats of the dog and cat families, we have relied to some extent on the fact that cats, large and small, may meticulously cover their dung by scraping together mud or debris with their forepaws, or place it within a scrape created by the hind feet. However, felines often leave scats unadorned, and dogs and coyotes at least, and possibly others in the group, are likely to give some vigorous backward scrapes at random with the hind legs. So even in the case of some of the dog family there may be some scratch marks nearby.

COYOTE

The coyote, *Canis latrans,* with its many varieties, known also as brush wolf and prairie wolf, is widespread and well known. However, coyotes east of the Mississippi are significantly larger than western counterparts.

I sometimes think that the most conspicuous coyote sign is his night song. Certainly a camp on the plains in the Southwest or in the western mountains is cozier when enhanced by the serenade of a coyote in the moonlight. Those who would follow the mammals in the wild should know something of the significance of this. Unaccustomed ears, trained by traditional journalism, might interpret the coyote voice as something doleful, a sad requiem that makes one crowd closer to the campfire. Or a flippant tongue might speak of the "yapping" of the coyotes.

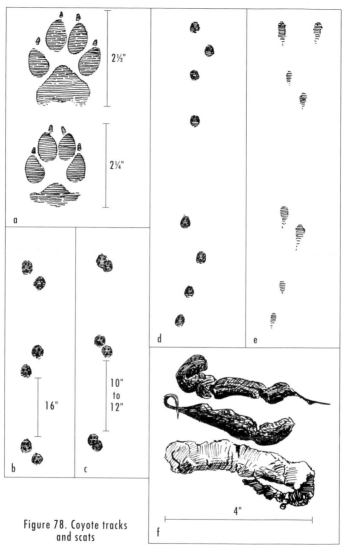

Figure 78. Coyote tracks
and scats

a. Tracks in mud: upper, front; lower, hind (Oklahoma).
b. Straddle trot.
c. Slow side trot.
d. Gallop.

e. Gallop into a bound.
f. Two types of scats.

But if the coyote could reflect and speak he would say this is his song, simply that. However it may appear to human ears, to the coyote it satisfies the universal impulse for expression of emotion, simple as that may sometimes be among the furred animals.

The coyote song is much higher pitched than that of the wolf. It is the soprano of this tribe. It may begin with a long clear call, breaking into a violent tremolo, these alternating irregularly. Or it may begin with a few barks that merge into the long call.

In our family we have used the coyote call as a signal. One evening as I approached within a quarter of a mile of camp I gave the coyote call to let them know in camp that I was coming. On an open hillside nearby a coyote replied with a lusty outburst!

Coyotes, and dogs that still retain this unsophisticated urge in their being, will often respond to such a call. Also coyotes may sometimes approach to investigate a simulated mouse squeak if they are within hearing, or the louder cry or scream of an animal in distress, if it can be imitated, and if you are in the lee of the coyote. In this instance, of course, it is the hunting instinct that is aroused.

When you happen to approach a coyote den, you may hear the parents bark, much like a dog.

The tracks follow the general pattern of the large wolf, and dogs, as shown in the accompanying sketches. As in the other Canidae, the front foot is larger than the hind foot. Front tracks of western coyotes measure 2¼–3¼ inches long (including claws) and 1½–2½ inches wide. Front tracks of eastern coyotes measure 2⅝–3½ inches long and 1⅝–2⅞ inches wide. The hind tracks of western animals are 2⅛–3 inches long and 1⅛–2 inches

Front (above) and hind tracks of a western coyote in Big Bend National Park, Texas.

wide, while eastern animals measure 2⅜–3¼ inches long and 1⅝–2⅜ inches wide. Note the differing patterns in the heel pads of the front and hind tracks in figure 78, a. This clear outline, however, rarely appears in the tracks you find. In loose snow and many other surfaces the heel-pad pattern is only suggested and usually appears roughly like that of the front foot. It is on firm snow, and mud, that this outline may appear.

One wintry day I followed a coyote trail in the snow for a short distance. Where the trail crossed an opening in the woods, the coyote had at several places turned aside to dig in the snow. Finally, at one place, he had been rewarded by finding the remains of a ruffed grouse. Some feathers were strewn about near the hole in the snow (see figure 80, a). At first I thought the coyote had pounced on a grouse, resting in its burrow under the snow, as grouse do, but the coyote had dug much too deeply for that, and there was no blood on the snow. So it would have to be the re-

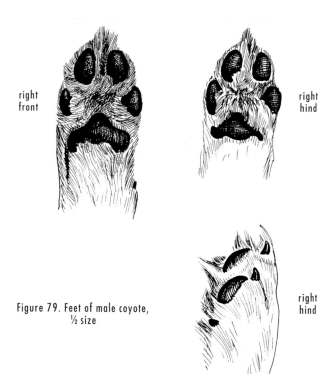

right front

right hind

Figure 79. Feet of male coyote, ½ size

right hind

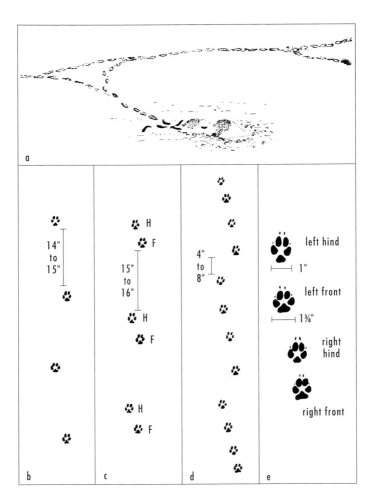

Figure 80. Coyote tracks in Wyoming

a. Coyote trail, showing excursions to one side of route to dig in the snow.
b. Direct-register trot, in which the hind tracks cover the front tracks.
c. Typical side trot.
d. A slow lope.
e. The lower series in d, to show details and order of front and hind feet.

mains of a bird lying far beneath the snow. Was it originally left there by the coyote for future use, or did the coyote's keen nose detect it under several feet of snow? I don't know.

This same coyote gave me the track patterns shown in figure 80: trotting in b, a side trot in c, and with a slow lope in d. The width of the trail, or straddle, was 4 inches in b and c, and 6 inches in d. In e we have an enlarged view of the lower track series of d. The hind pad of the hind foot generally registers in a somewhat circular form in snow, and smaller than the well-lobed front pad.

In the track patterns b, c, d, and e of figure 78, the hind prints are forward of the front prints. When the coyote was galloping fast the leaps shown here measured 32 to 61 inches, in snow. In other cases they leaped 51–120 inches.

Coyote scats are extremely variable in size. The residue from pure meat is likely to be semiliquid. The scats consisting of much hair are likely to be large. Those resulting from a diet of pine nuts or chokecherries are likely to crumble. In size, coyote scats overlap those of the wolf and the red fox, and those of pups are of course much smaller.

Scats are likely to be deposited along trails. In the western mountains they will be found where a trail comes over a little knoll, or level place, or any other point of special interest to the coyote. There is likely to be an accumulation where two trails cross — as when a ridge trail dips down a slope and crosses a trail coming over a saddle.

Such accumulations may be at the scent post used by dogs, wolves, and foxes. This may be a rock, a prominent tuft of grass, a stump, or almost anything that stands out, where urine is deposited by members of the dog family who casually investigate it when passing by.

GRAY WOLF AND RED WOLF

The gray wolf, *Canis lupus,* once ranged widely over North America and far into the Arctic. Now this glamorous animal has been nearly crowded out of the United States. It is still to be found in small numbers in northern Minnesota, Wisconsin, and Michigan, but occurs more commonly in Canada and Alaska, where it reaches a large size. More recently, the gray wolf has reclaimed parts of Montana, Idaho, and Washington, and wolves have been successfully reintroduced to central Idaho, Yellowstone National Park, and the Arizona–New Mexico border, from which their ranges have already begun to expand.

The wolf of far eastern Canada and Algonquin National Park, an animal that has occasionally slipped into northern New England,

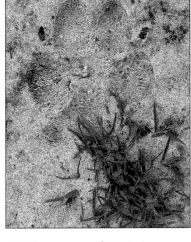

The front track of a red wolf in coastal North Carolina.

is now considered by many to be a distinct species and is called the eastern timber wolf, *C. lyacon*. There is still debate as to whether this animal is truly a distinct species from the gray wolf. In tracks and signs it is a much smaller animal than western relatives, and trail measurements tend to be just a bit larger than red wolves.

The smaller red wolf, *C. rufus*, may have remnant populations in Texas, parts of Loui-siana, Arkansas, Missouri, Tennessee, and northern Mexico. Yet the genetic stronghold for the red wolf is coastal North Carolina. This little wolf approaches the coyote in size, at least in some specimens.

In these times of disappearing wilderness, to have heard the howl of the wolf in wild country is a memory to be cherished. I recall seeing wolf sign in central Labrador many years ago, in company with an Indian. Even then the wolves were becoming scarce in that country, and we were impressed by the fact that a wolf had been at the spot where we were standing.

One night four of us, including our year-old baby, were encamped on a gravel bar of the Porcupine River, in northeastern Alaska. It was clear September weather, and we slept that night in the open without tent. At dawn we were awakened by a voice across the river. Soon we realized that we were being serenaded by two wolves, one upstream, the other below our camp. First one, then the other, raised its muzzle and howled. Apparently we were intruding on their home ground. At any rate, we lay there in the crisp autumn morning, comfortable in our sleeping bags, and listened to this song of the Arctic wilderness with a feeling of awe.

The wolf song is a long monotone, lacking the "yapping" and tremolo of the coyote. It is very similar to the howling of the Alaskan malamute and other deep-chested breeds.

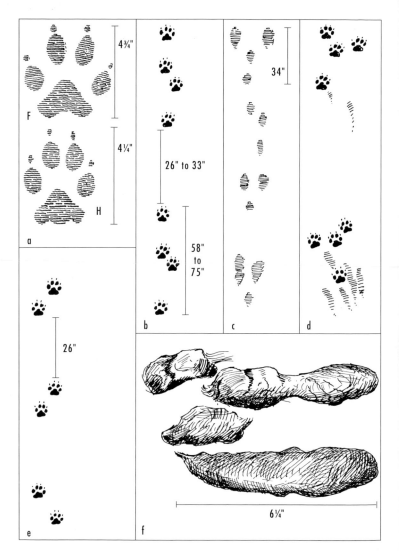

Figure 81. Gray wolf tracks and droppings

a. Tracks in mud (Alaska).
b. Lope.
c. Irregular run.
d. Bounding gait.
e. A fast trot.
f. Droppings (northern Minnesota).

The tracks of a walking gray wolf in Yellowstone National Park. Look carefully for the tracks of the coyote moving in the near opposite direction, and the numerous tracks of Canada geese.

For further contemplation of the probable significance of the wolf howl and its ceremonial actions, refer to chapter 2 of the monograph *The Wolves of Mount McKinley,* by Adolph Murie.

The tracks figured here are those of the large Alaskan variety. The front foot is larger than the hind foot and in mud or wet sand will measure 3¾–5¾ inches long (with the claws) by 2⅞ to over 5 inches wide, depending on speed and the spread of the toes. The front toes tend to spread much more than the hind ones. The hind tracks measured were 3¾ – 5¼ inches long by 2⅝ – 4½ inches wide. In snow the tracks tend to be somewhat larger, the front track reaching 6 inches in length. The largest track measured was one in wet sand, on the Porcupine River, Alaska, which with toes widely spread measured slightly over 6½ inches in length and width. In figure 81, a, note the difference in the outline of the front and rear foot pads. A similar difference can be seen in the dog tracks, figure 75.

Figure 81 illustrates various gaits, which are quite like those of a dog of corresponding size. In fact, in the North wolf tracks can easily be confused with those of Alaskan malamutes.

Wolf droppings are illustrated here, but there is such variation in size that many could be confused with those of the dog, as well as with larger ones of the coyote.

ARCTIC FOX

The Arctic fox, *Vulpes lagopus,* ranges throughout the Arctic regions, coming as far south as the Aleutian Islands in Alaska. It has

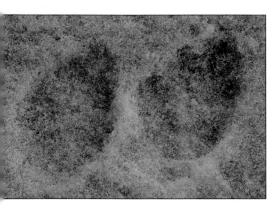

These Arctic fox tracks in Churchill, Manitoba, are minutes old, yet the toes are barely evident due to the copious fur which covers their feet in winter.

two winter color phases, white and blue. The majority become white in winter, but some members of the family turn a bluish gray, from which the furriers derive the term "blue fox." All Arctic foxes are brownish in the summer.

We have here a fox that is far different from the traditional Reynard of the fairy tales or the sly fox of numerous stories. One day I was sitting on a knoll on one of the Aleutian Islands, watching a pair of blue foxes below me on the beach. One of them spied me and, to my surprise, came charging all the way up the hill, bumped me in the knee with its muzzle, and hurried back to the beach. History tells us that when Vitus Bering camped in the Commander Islands, where these foxes were unusually plentiful, the animals would come into the tents and would sometimes nip the men resting there. In Greenland, too, the Arctic foxes live near human habitations. Accordingly, except where they are intensively trapped and therefore frightened by men, you may expect to find the Arctic fox easy to approach and to observe.

The feet of this fox are hairy underneath, especially in winter, more so than those of other foxes. However, an orthodox canine

Figure 82 (opposite)
a. Tracks in sand: upper, front foot; lower, hind (Aleutians). Some tracks in sand were as much as 2¾ by 2½ in.
b. The common slow loping gait used by exploring Arctic foxes.
c. Fast gallop track pattern.
d. Well-furred left hind foot, in winter (Hudson Bay, 1915).
e. Arctic fox droppings; those containing feathers (2 upper samples) were stringy and thin.

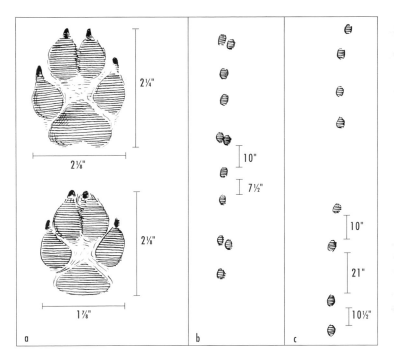

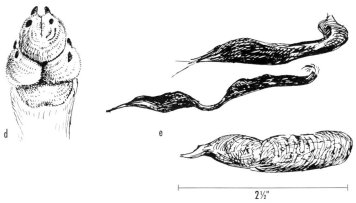

Figure 82. Arctic fox

track, with impression of toes and heel pad, is produced in spite of the hair, particularly in summer when the hair is less dense. When the imprints are clear the track of the Arctic fox may be distinguished from that of the red fox by the absence of the barlike impression that appears in the heel pad of the latter. It may be difficult to distinguish this feature in light or loose snow, or in other unfavorable surfaces. Front tracks measure 2¼–2¾ inches long and 1¾–2⅛ inches wide. Hind tracks are 2–2⅜ inches long and 1⅝–1⅞ inches wide.

In general, the droppings are similar to those of the red fox, but often, when feeding on crustaceans, the Arctic fox will leave droppings pinkish in color and tending to bleach out white. In the Aleutian Islands I occasionally found gull castings that somewhat resembled blue fox droppings, especially when the bird had fed on crustaceans.

The call is a raucous cry or bark, a rather harsh sound.

KIT AND SWIFT FOXES

The kit fox, *Vulpes macrotis,* is a desert animal found in the arid Southwest and in northern Mexico. The swift fox, *V. velox,* is a creature of the plains regions of eastern Colorado north through the Dakotas into southernmost Canada, though today populations are small and scattered. It has been a long debate as to whether these two closely related foxes are truly different species, or variations of the same animal. Today it is believed they are two distinct animals, though their sign is very similar.

These trim graceful creatures are the smallest of our foxes, and the trail is correspondingly dainty. Kit fox tracks measure:

front, 1–1¾ in. long, ¹⁵⁄₁₆–1½ in. wide
hind, 1⅛–1⅝ in. long, ⅞–1¼ in. wide

When dealing with such close distinctions in the tracks from relatively small feet, it must be kept in mind that there will be

Figure 83 (opposite)

a. Tracks of gray fox.
b. Tracks of kit fox, in snow.
c. Feet of gray fox.
d. Track of kit fox, in dust.
e. Track of kit fox, in loose sand.
f. Trotting trail of gray fox.
g. Kit fox trail, created by mixed trotting gaits.
h. Tracks of kit fox when galloping.
i. Droppings of kit fox or gray fox (Texas). These are often indistinguishable.

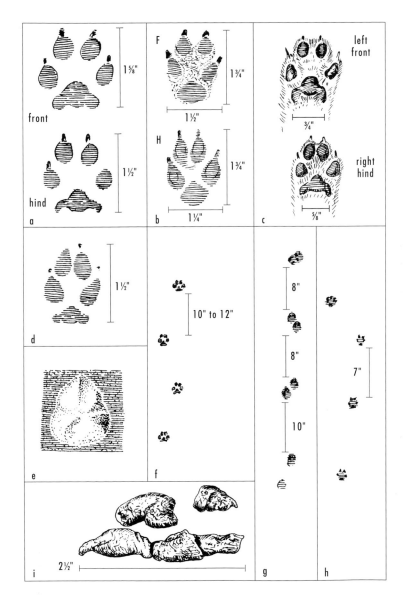

Figure 83. Gray fox and kit fox

Perfect, dainty kit fox tracks in mud in Death Valley National Park, California.

considerable differences of measurements on different surfaces, whether it be loose or firm sand, wet or dry, or snow. In loose sand the details of kit fox tracks are blurred. Therefore, in sand dune country of the Southwest, you will find a kit fox trail composed of a line of depressions in which sand has slid down to conceal the toe marks (figure 83, e), and you will only be able to identify it by the size of the footprint as a whole and the track pattern in general. Consider also to what extent fur covers the bottom of the feet. Some kit and swift foxes have fur nearly covering all their toe and palm pads, which protects them from hot sands, while other animals may show far more pad and create tracks more similar to gray foxes.

The kit fox scats (see figure 83, i) are similar to those of the gray fox. If such droppings are found in a strictly desert area, they are most likely to be those of the kit fox. Should scats contain any substantial fruit they can also often be attributed to gray foxes, as kit and swift foxes are the most carnivorous of the foxes.

The burrow is often out on the sandy plain and may have several entrances, which measure approximately 8 inches in diameter, sometimes larger. The kit and swift foxes use their dens as a refuge the year round to protect themselves from midday temperatures and sun, as well as to provide cover from potential predators. In the Mojave Desert, kit fox den entrances were surrounded by discarded kangaroo rat tails, their principal prey in the region.

I have never heard the voice of the kit fox. Ernest Thompson Seton has described it as a bark much like that of the red fox, but on a smaller scale.

The red fox, *Vulpes vulpes*, is one of the best-known characters in history and legend, widely spread over the temperate and northern regions of the world. For its combination of beauty and grace and intelligence it has had the attention of artists, poets, and naturalists, and merits the attention of those who would read the signs of the out-of-doors.

On January 30, 1841, Henry Thoreau wrote in his journal:

> Here is the distinct trail of a fox stretching a quarter of a mile across the pond. Now I am curious to know what had determined its graceful curvatures, its greater or less spaces and distinctness, and how surely they were coincident with the fluctuations of some mind, why they now lead me two steps to the right, and then three to the left. If these things are not to be called up and accounted for in the Lamb's Book of Life, I shall set them down for careless accountants. Here was one expression of the divine mind this morning. The pond was his journal, and last night's snow made a *tabula rasa* for him. I know which way a mind wended this morning, what horizon it faced, by the setting of these tracks; whether it moved slowly or rapidly, by the greater or less intervals and distinctness, for the swiftest step leaves yet a lasting trace.

The red fox has a variety of calls, variously described as squalling, screaming, and barking, and of course each has its own significance. I do not know the purpose of the red fox bark, unless it may correspond to the howl of the coyote. At least on one occasion, when three of us who were boys in Minnesota came

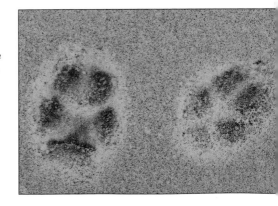

Fresh red fox tracks in southern New Hampshire appear blurred by the heavy fur covering their feet. Note the characteristic chevron-shaped line which appears in the palm pad of the front track.

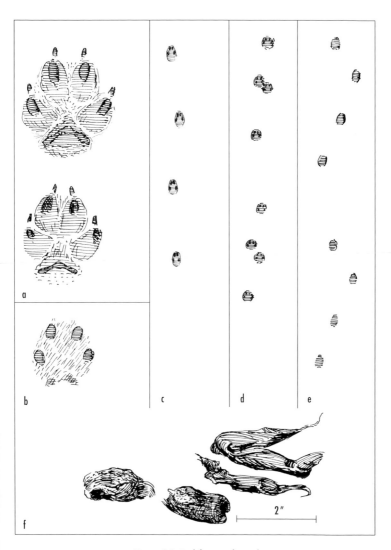

Figure 84. Red fox tracks and scats
a. Tracks in mud: upper, front; lower, hind (Alaska Peninsula).
b. Track on firm sand (Alaska).
c. Trotting gait.
d. Loping gait

e. Gallop, with hind tracks
   out in front.
f. Scats.

upon a fox den in the woods, the outcry from the parent foxes was clearly a protest, an expression of alarm at our intrusion. It was a combination of a sharp bark and scream, harsh and penetrating. This was quite different from—more shrill than—the regular bark.

The trail of the red fox is traditionally known as a line of dainty footprints in an almost straight line. This is largely true; but, as in the trails of other animals, the pattern varies greatly with the gait and speed. In the almost "straight line" of the trotting pattern the trail may be 2–3¾ inches wide. The width of the coyote trails in similar patterns is from 2¼ to 5¼ inches.

When contemplating the "dainty" form of the red fox tracks, note also the difference between the wide, sprawling front-foot print as contrasted with the narrower, more pointed hind-foot print. Some prints, especially in a shallow, firm snow or similar medium, show only the portion of the toes and heel pad that protrudes from the hair (figure 84, b). In such prints the heel pad appears to lie far behind the toes, without the lobe extending up between the two hind toes as in dogs and wolves. However, in deeper snow and in soft mud, the entire surface of the pads, including the portion covered with hair, makes an impression, and then we get a track (figure 84, a) quite comparable to that of a small dog or coyote, or to other foxes.

The red fox track has one good characteristic that is distinctive, if you have a track showing details. The heel pad has a chevron-shaped or straight "bar" protruding from the hair of the foot, as shown in figure 84, a. In mud, shallow snow, or otherwise a firm surface, this bar may show without the rest of the pad. Then you get a print something like figure 84, b, with the whole bar, or only two ends. In deeper material the bar makes a little groove in the bottom of the heel-pad print. In many parts of the country, they are the only canine with nearly completely fur-covered feet. Front tracks measure 1⅞–2⅞ inches long and 1⅜–2⅛ inches wide. Hind tracks are 1⅝–2½ inches long and 1¼–1⅞ inches wide.

Red fox scats vary greatly in size and appearance, depending on the quantity and kind of food eaten. The samples illustrated in figure 84 are representative, showing that they have the general form of the scats of any of the Canidae. The variation in size, moreover, makes it difficult to identify some samples, for they may be confused with those of the gray fox in the eastern states, or with the Arctic fox in some parts of Alaska, not to mention the coyote in the lands in between. Normally, however, coyote scats will be larger.

Foxes dig their dens in a variety of places; in the woodlands of the eastern states or on the open plains of North Dakota. They have been known to make their home in a hollow log, and sometimes they excavate an old woodchuck burrow. Such dens may be identified by tracks in the neighborhood, or by fox hairs clinging to the entrance. There is also a distinctive scent, which is especially apparent in mating season near urine scent posts. Francis H. Alien once told me: "I have often smelled the fox scent at some distance from any den. It resembles skunk but has a different quality and is not so strong." Scent posts and dens can be tracked with the nose as much as several hundred yards on a windy day, and by those who can smell better than I, much further.

COMMON GRAY FOX

The common gray fox, *Urocyon cinereoargenteus*, with its related species and subspecies, ranges throughout most of the United States except the north-central and northwestern states (including the Dakotas, Nebraska, Wyoming, Montana, Idaho, most of Washington, and parts of adjacent states). These areas are either open plains country or high mountain country, and such environment is shunned by the gray fox. Its chosen country is the woodland of New England, the brushlands of the Southwest, the chaparral and woods of the Pacific Coast, and the congenial environment down through Mexico and Central America. The Is-

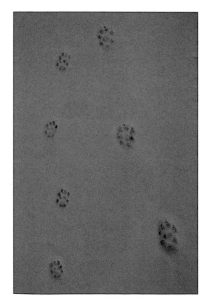

Compare the trotting trail of a gray fox, on the right, and the walking trail of a feral cat along the snow-covered Connecticut River between Brattleboro, Vermont, and Hinsdale, New Hampshire.

land gray fox, *U. littoralis,* is slightly smaller, very endangered, and inhabits the Channel Islands off southern California.

In the Southwest, the gray and kit foxes often occupy the same general areas. And here we have some difficulty in distinguishing their signs.

Common gray fox tracks measure:

front, 1 ¼–1 ⅞ in. long, 1 ³⁄₁₆–1 ¾ in. wide

hind, 1 ⅛–1 ¾ in. long, 1 –1 ½ in. wide

However, the Island gray fox may be smaller.

Tracks are imperfect, and it is to be noted that where the imprints of the gray fox feet are not detailed, the heel pad of the hind track may appear chiefly round and small, the lateral portions failing to register as readily as in the heavier front track. There is the same tendency to be found in the coyote tracks.

It should be remembered that the gray fox is the only fox that climbs trees readily, especially leaning trees, a fact that may possibly be revealed in its trail at times.

I have not been able to distinguish the droppings of the gray and kit foxes. Those shown in figure 83, i, could be either. Such small droppings in the eastern states, where the kit fox is not found, would of course be those of the gray fox.

The dens may be in any of a great variety of places, in the ground, among rocks, a hollow tree, a woodpile, or even in a drainage pipe. When the pups begin to come above ground the front of the den will be stomped down from continuous play, wrestling, and feeding. Small scats will litter the area and the stench of decaying meat will be heavy in the air.

The most common voice of the gray fox is an unusual bark, a rough, guttural, raspy sound, which might not be readily recognized as a bark at all. I've also been awoken by growling gray foxes circling me, and I've heard a variety of growls and grunts from young foxes defending food portions or establishing dominance within the litter.

Family Ursidae: Bears

BLACK AND BROWN (GRIZZLY) BEARS

THE COASTAL BROWN BEAR and interior grizzly are now considered one species, *Ursus arctos,* and inhabit Alaska, western Canada, western Montana, the Yellowstone Park region, and portions of Idaho and Washington. The black bear, *U. americanus,* exists in wooded areas through much of the West and North, and in portions of the East including the Appalachian regions and Florida. These two bears are discussed together because there are so many similarities.

In the spring of 1921 I was camping on Robertson River in the Alaska Range. One morning when I walked over to the river for a pail of water I found grizzly tracks in the sand. They led up to within 5 yards of my tent, where the bear had been sniffing about during the night. I had plaster of Paris with me and took impressions of the tracks, almost the first track casts I had ever made, which are shown in figure 85, b. These tracks illustrate the

Front and hind tracks of a coastal brown bear along the central Alaska coast.

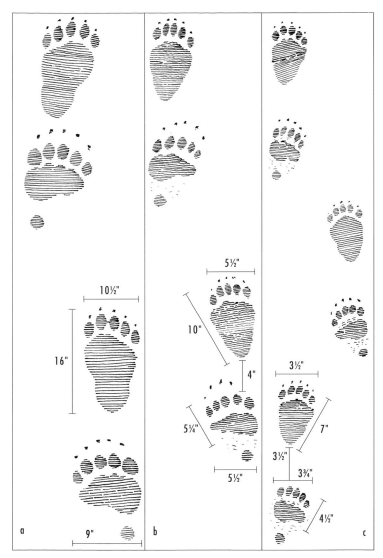

Figure 85. Walking gaits of bears
a. Coastal brown bear. b. Interior grizzly (brown bear) c. American black bear.
All three bear trails shown here are created by walking gaits in which the hind
feet overstep the front tracks on a given side of the body.

long front claws characterizing the grizzly, in contrast to the shorter ones of the black bear (figure 85, c). The incident of finding the tracks shows how much can be added to outdoor experience by recognition of animal signs.

To illustrate something else I shall tell of another experience in the same part of Alaska. Prospectors in the Tanana River region spoke of the "glacier bear" and said you could distinguish the tracks, which are "halfway in size between grizzly and black bear tracks." I found this bear, which proved to be a small-sized whitish grizzly, not the true "glacier bear." And indeed it really was possible to distinguish its tracks in many cases. It was definitely smaller than the normal large grizzly whose cast I had obtained. Was this then a different species? Different sex? I still do not know. This experience emphasizes the importance of allowing for deviation in size of tracks due to sex, age, individual variation, as well as condition of mud, sand, or snow. Remember, too, that the round heel pad of the front foot, shown in these perfect tracks, very often does not register.

The trails of figure 85 show the typical walking gait, hind foot a little forward of the front foot on the same side. Observations on a cub grizzly showed that when walking very slowly the hind foot registered exactly in the front-foot track, though when stepping out more vigorously the hind foot overstepped the front-foot track to produce the pattern here described.

These variations in track pattern are shown in figure 86, d. Here the female black bear placed the hind foot squarely in the front print (in the lower part of the figure), then began to overstep the front track (in the upper part) to produce the pattern so often seen. Note that in c the cub did not let down the heel of the hind foot, so all of these footprints resemble front tracks.

The tracks of a black bear in an overstep walk in cental Alaska.

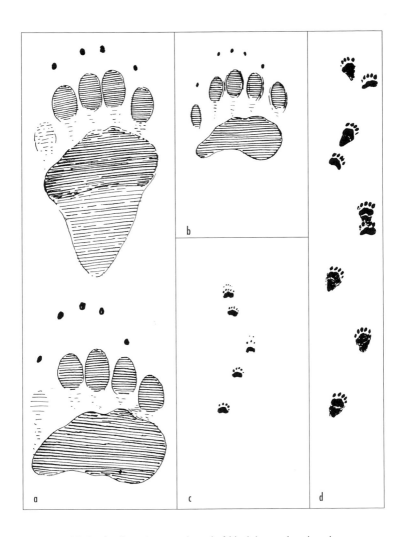

Figure 86. Tracks of yearling grizzly and of black bear cub and mother

a. Yearling grizzly: hind track, 6½ in. long by 3¾ in. wide; front, 4 in. by 4 in.
b. Cub's front track, 2½ in. both length and width.
c. Trail of cub, in erratic pattern.
d. Trail of mother, changing from a direct-register walk, where hind tracks cover front tracks, to an overstep walk, where hind tracks lie beyond front tracks.

Notice, too, that the "big" toe of the bear foot is the *outer* one, not the inner one of the human foot. You will find in bear tracking that in dust or shallow mud quite often the "little" inside toe leaves no mark, so the footprint appears to be four-toed. Or the inner toe may be only faintly seen (figure 86, a).

In figure 87 we have the grizzly gallop in a, black bear gallop in e, with hind feet out in front with each leap. These show two gallop patterns that, with still other variations, may be made by any of these bears.

All bear groups have well-established trails. Those of the Alaska brown bear, especially on the open tundra of Alaska Peninsula, are striking. They take several forms. In marshy tide flats or along the salmon streams, they appear like any simple trail through heavy vegetation. On the drier tundra of the uplands they take the form of two parallel ruts, made by the right and left feet, where bears have traveled for years (figure 87, c). I have walked in such bear trails but found it awkward, for the brown bear's hips and shoulders are much greater than a man's, and I found it necessary to spraddle widely to keep in the ruts.

There is still another type of bear trail, shown in figure 87, d. Sometimes big brown bears and black bears get the habit of stepping repeatedly in the same footprints, until a series of pits in zigzag fashion is developed. In order to travel in brown bear trails I found it necessary to make a half hop from one to the other. It is interesting to note that on mossy tundra this type of trail, if abandoned for some years, grows up in grass because the seeds find bare soil available in the bottom of the pits. So you may find a zigzag line of grass clumps marking an old brown bear trail. In the trails of both the black bear and the grizzly you will also occasionally find a tendency toward the formation of zigzag pits in locations where many bears have stepped in each other's footsteps for years. In the Northeast, black bear trails of this sort are most often near wetlands and refuse dumps.

During June, and on through the summer, the salmon ascend the streams to spawn. Then the brown bears come into the lowlands to feed on the salmon. You find their trails, well worn, along

Figure 87 (opposite)
a. Grizzly trail, fast gallop, in snow (Yellowstone National Park).
b. Grizzly tracks, slow lope, in snow (Alaska).
c. Double-rutted trail of Alaska brown bear on high tundra.
d. Staggered prints of Alaska brown bears, each bear stepping in the same place repeatedly over a long period of time.
e. Fast bound of a black bear.

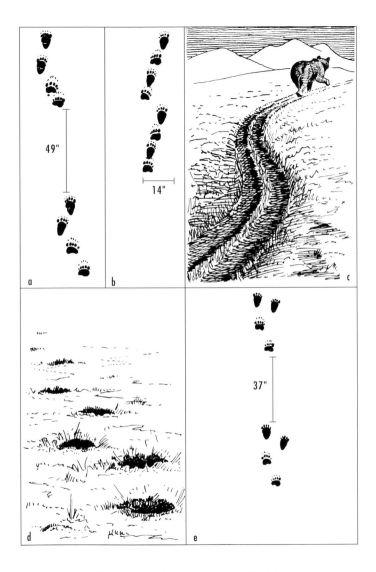

Figure 87. Grizzly, Alaska brown bear, and black bear trails

the banks of the streams, and you find resting places scooped out in the vegetation among the willows and alders. Sometimes there are beds specially prepared by scraping together a pile of moss, perhaps a dozen feet in diameter, apparently to serve as a mattress. Once when I found some of these piles of vegetation I wondered if they had served to cover some stored food, but did not find evidence of it. Rather, these seemed to serve as comfortable mossy mattresses. Black bears are also known to construct mattresses upon which to bed when temperatures dip below freezing. They are often formed of branches and leaves torn from nearby bushes.

In wooded country it will be noticed that a bear trail will go under obstructions that an elk, for example, would have to go around.

Then there are the bear trees. Bears will bite and pull off strips of bark from the trunks of pine, spruce, and fir trees in spring and early summer, sometimes girdling the trees. Having pulled away the bark, they will scrape off the juicy substance on the wood with their incisor teeth, leaving vertical tooth marks (figure 88, a). I tasted some of this. At first there was syrupy sweetness, followed immediately by turpentine! But the bears like it.

Black bears will also climb aspen, beech, and other tree species to feed on nuts, fruits, or spring leaf shoots. The claw marks re-

Figure 88. Bear trees
a. Tree stripped for the juicy pulp under the bark.
b. Old healed scars on an aspen, where a bear had climbed.
c. A grizzly rubbing tree (Alaska).

URSIDAE

main in the soft smooth bark and a scab forms on each, so that the climb remains recorded for the life of the tree (figure 88, b).

There is still a third tree. Bears like to rub themselves, usually on a tree, but occasionally on a bush or stump. They will rub and rub, sometimes grasping the tree and clawing it, sometimes biting it as they stand on their hind legs. Often this tree is in a prominent place, on a point or beside the trail, where it easily comes to the notice of the bear, and it is rubbed and scratched repeatedly until we recognize it as an established "bear tree" (figure 88, c, and figure 89, a). Generally pitch oozes out and you will find hairs stuck in it or clinging to the bark. Such hairs should not be confused with buffalo hairs in places like Yellowstone Park, for instance, since the buffalo also rubs trees (see page 311). This type of bear tree has been construed as a signal tree and certainly conveys information including sexual status and dominance. Undoubtedly it serves as a "sign" post, similar to the scent post of the dog tribe, but it is also a place for comfortable rubbing. In national forests or in national parks you will often find trail signs chewed up by bears, which is also a marking behavior. All bear species will also straddle vegetation and urinate on it as they walk

Figure 89. Bear trees and black bears foraging
a. A tree clawed by a black bear (Great Smoky Mountains, Tennessee).
b. Black bear tearing open a rotten log for insects. An anthill in the foreground has been scooped out for the same purpose.

A log opened by a black bear looking for insect larvae in Montana.

over it. The saplings spring up behind them and serve as yet another scent marking behavior and sign post.

Bear droppings may be confusing at times, though generally they have a distinctive form and a tendency to maintain a fairly even diameter (figure 90). Bears will eat meat whenever they can, killing animals as large as a moose—or feeding on carrion. On such a diet the scats are tubular and consist chiefly of hair, or may be flat and amorphous if containing mostly the remains of meat and internal organs. Remember, however, that a bear is pretty much of a vegetarian, and a large proportion, probably the majority, of the scats you find will consist of grass, roots, or fruits. There may be a mass of wood debris mixed with ants, or a mass of pine nuts, or berries in season. Coyotes eat pine nuts too, but their droppings are smaller. A straight diet of strawberries may produce a semiliquid mass. The same applies to a diet of fish, as with the big brown bear of southern Alaska. Grizzly and brown bear droppings are generally larger than those of the black bear, but they do overlap in size.

There are other feeding signs. You may find an old log turned over or torn apart where a bear has been looking for beetles, ants, or more likely insect larvae (see figure 89, b). Rocks are turned over for the same purpose. Anthills are scooped out, so that the swarming ants can be licked up in quantity. In Yellowstone Park I found spots in meadows torn up where a bear had scented an underground ant colony, and in the same park my brother found them turning over buffalo chips for beetles. Once in Alaska I found a bank swallow nest clawed out of a bank by a black bear, the claw marks being obvious. Bears dig for plant roots, and sometimes their diggings are very prominent, as if someone had

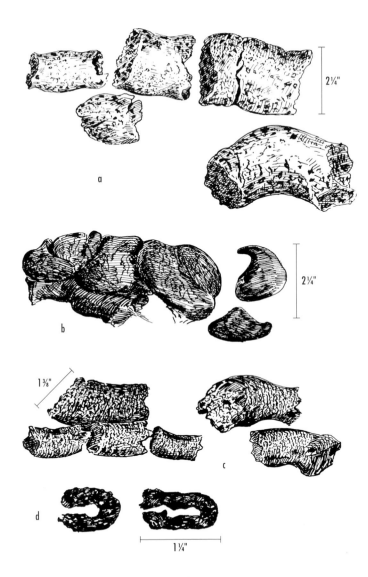

Figure 90. Bear droppings
a and b. Grizzly scats.
c and d. Black bear scats, d being that of a cub.

been sporadically digging a garden plot. Or you may find an excavation where a bear has dug a ground squirrel from its burrow, a pocket gopher's root cache, or a ground hornet's nest.

Occasionally you might find a food cache where a bear has covered part of a carcass of a deer or other animal for future use. This is very similar to the cache of the mountain lion, and may be difficult to distinguish. Look for tracks or some other sign of the animal nearby, but exercise great caution as both bears and cougars are known to defend such caches from intruders.

Dens for hibernation are not often found. They may be in a hollow under an upturned root, in the case of black bears, or in any of many natural cavities. On Unimak Island, Alaska, I found big brown bear dens in long underground tunnels which were natural cavities in the lava beds. In interior Alaska I found grizzly dens excavated in the base of a hill. One day my brother showed me a den, excavated in August, that he had found high on a mountain slope. We crawled inside to see what it was like, feeling quite certain that the bear was not at home. (See figure 91.)

Bears leave plenty of sign, and its interpretation furnishes a good story. They also produce sound to express a variety of emotions. If you find yourself near a bear, whether black or brown, and he makes a coughing sound or "chops" his jaws, look out! The bear is in a threatening or surly mood. Sometimes when the animal is more mildly annoyed, it will growl in a low, smooth-voiced manner that is hard to describe. The most pitiful sound I have

Figure 91. Grizzly bear den on a high slope (Denali National Park, Alaska)

heard in nature has come from a bear in great trouble, as when it is wounded. It is a strong, variable, moaning sound, so realistically human in its quality that it is heartrending to listen to. The bears, then, express themselves with a considerable variety of growls, coughs, sniffs, and certain whining or bawling sounds. Ordinarily, however, as you find these animals in the field, you are not likely to hear their voices.

POLAR BEAR

Few of us have the chance to trail the polar bear, *Ursus maritimus.* You must travel far to the north to visit the polar region. Churchill, on Hudson Bay, is famous for its polar bear population, which congregates each fall to await the freezing of the bay and the start of a new seal hunting season. There the male bears lounge about in wait, wrestling in short bouts of play and in preparation for real fights during the mating season to come. The footprint shown in figure 92, b, was obtained in the Woodland Park Zoo in Seattle, where a polar bear was induced to step in moist sand. The polar bear soles are extremely hairy, hence the various pads are not clear-cut. Figure 92, d, from a photograph, shows the overall shape of some tracks but does not reveal details of foot structure. The illustrations in figure 92, a, are the front and hind tracks of a large male in shallow snow near Churchill, Manitoba.

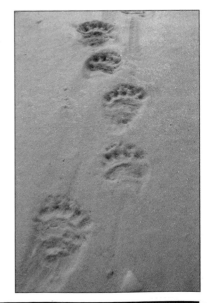

Any bear tracks that you are likely to see on the ice of the polar sea or on the islands of the Arctic coast or the Bering Sea, should be those of the polar bear.

The droppings of this bear consist of vegetation or remains of seal, fish, or other carnivorous fare. Contrary to traditional natural history, the polar bear has been found to feed on vegetation as well as meat, specifically in the Bering Sea and on the Arctic islands.

The walking trail of a subadult polar bear in light snow near Churchill, Manitoba.

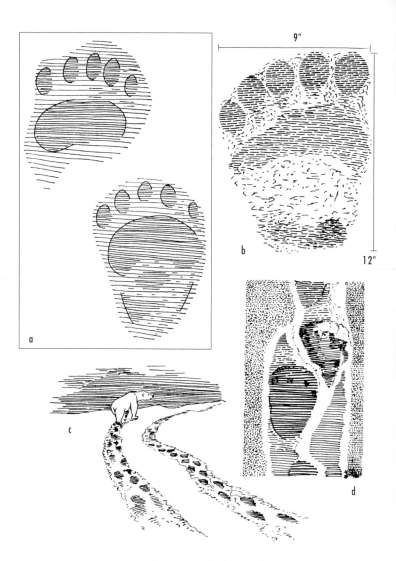

9"

12"

Figure 92. Polar bear tracks
a. Front and hind tracks of a large male in shallow snow (Churchill, Manitoba).
b. In sand (Woodland Park Zoo, Seattle).
c. Snow trails of walking bears.
d. In snow (Arctic island, Canada, from photograph by C. O. Handley Jr.).

FAMILY OTARIIDAE: HAIR SEAL;
FAMILY PHOCIDAE: EARED SEALS

AQUATIC ANIMALS like seals and sea lions naturally do not leave traces except when they come ashore. The common harbor seal, *Phoca vitulina,* hitches along awkwardly on land. If this happens in mud or sand, the drag marks of the body may be seen, as well as the holes made by the nails of the front flippers (see figure 93, a). If the seal hauls out on ice or another hard surface covered with snow, the scrape marks appear as in figure 93, b.

Hair seals generally deposit scats in the water, but when they are found on land they appear claylike or pasty, about 1 ½ inches in diameter.

On Arctic ice floes seals are readily discovered in the distance, since their dark color contrasts vividly with the snow and ice. They have the habit of sleeping intermittently, raising their heads to look around every few minutes. Eskimos and Inuits often approach a seal by crawling up during the intervals when the head is down, timing it so that they are mo-

The trail of a harbor seal in deep, soft sand in central Oregon.

tionless when the seal is on the lookout. On the winter ice the seals keep open breathing holes in the ice. When the snow forms a mantle on the ice, the seal comes out of the hole but not through the snow, thus virtually forming an igloo for itself under the snow crust.

A sea lion rookery is easily detected at some distance by the raucous roaring, and at a closer distance by the strong smell of ammonia. Sea lion droppings are similar in substance to those of the seal, but larger, about 2–2½ inches in diameter, depending on sex and age (see figure 93, c). They may also be soft and formless.

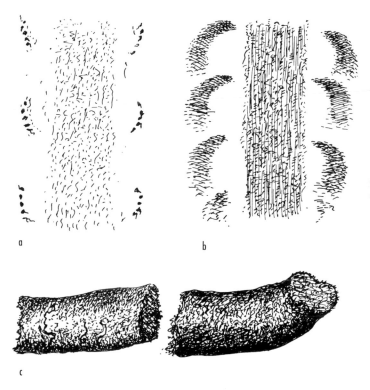

Figure 93. Seal tracks and scat

a. Tracks of harbor seal in mud (after photograph by Dr. Victor B. Scheffer).

b. Tracks of seal in ½ in. of snow (National Zoological Park, Washington, D.C.).

c. Droppings of Steller's sea lion, *Eumetopias jubatus,* diam. about 2 in. (Aleutians).

FAMILY PROCYONIDAE: RACCOONS, RINGTAILS, AND COATIS

HERE IS an interesting little group in our animal world. Its members inhabit the southern portion of our continent, though the raccoon finds its way north into southern Canada.

The tracks of this group tend to be plantigrade, or flat-footed, like those of the bears, only in miniature. The ringtail cat, however, does not follow this pattern. Figure 95 shows the tracks of this group, drawn to scale for comparison.

RINGTAIL

The secretive ringtail, *Bassariscus astutus,* known also as cacomistle in Mexico, is a warm-climate animal of Central America and Mexico. In the United States it has found congenial habitat in the West and Southwest—southern Oregon, California, southern Nevada, Arizona, parts of Utah and Colorado, and parts of New Mexico and Texas.

The crisp trail of a walking ringtail in deep mud in Sacramento Valley, California. In this case it is an understep gait.

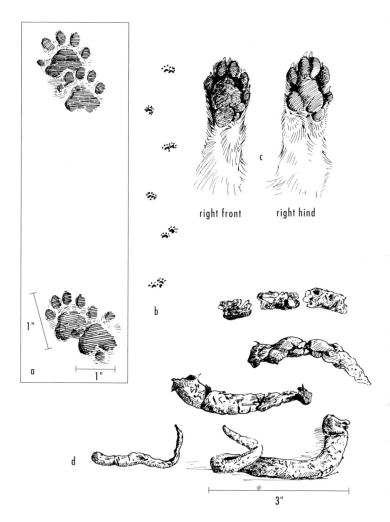

right front   right hind

c

b

a

d

Figure 94. Ringtail

a. Typical tracks, in soft dirt, hind track partly covering front track.

b. The trail pattern.

c. Feet (after Seton).

d. Scats, showing great differences in size and shape. Upper sample, from Texas (1950), contains insects; other samples, from New Mexico (1939), contain mammal remains.

The ringtail is strictly nocturnal and seldom seen, and in its dry habitat the tracks are not readily found. However, in Cottonwood Cave, New Mexico, in 1939, by the light of a candle I found many of its footprints in the loose dirt on the rock ledges, and some of the scats shown here. In this cave the ringtail had been feeding on bats almost exclusively. I also often find tracks near water sources such as rivers and water holes, especially where water is adjacent to ample cover afforded by cliffs, large boulders, or forest.

The tracks and trail patterns are quite catlike. Although this is one of the animals that have five toes on both front and hind feet, it is quite common for only four to register, especially in hind tracks. However, the front tracks of ringtails almost always show five toes and very often show the characteristic round pad at the posterior edge of the track. These facts and the larger palm pad should help distinguish their tracks from those of small feral cats. Front tracks of ringtails measure 1⅛–1⁷⁄₁₆ inches long and ¹⁵⁄₁₆–1¼ inches wide. Hind tracks are 1–1⅜ inches long and ⅞–1¹⁄₁₆ inches wide.

Their diet is quite varied and includes rodents, bats, insects, and fruit. Consequently the scats vary. In Texas, where miscellaneous food was eaten, the scats were often found broken up in short lengths, and crumbled easily when dry (see upper sample, figure 94, d).

The ringtail finds shelter, and makes its nest, in hollow trees, cavities in rock piles or cliffs, and in caves.

Little seems to have been reported on the voice of the ringtail, although it is known to produce a bark like that of a small dog. I have never heard it.

ORTHERN RACCOON

The well-known northern raccoon, *Procyon lotor,* may be found from the Atlantic to the Pacific, from the southern states to lower Canada. It generally avoids the dry desert, but I saw its handlike tracks at a watering hole among the cacti in southern Texas. On the sandy beaches of the Pacific Coast you may find it beachcombing.

The track is distinctive and easily identified, as shown in figure 96. It has five toes on front and hind feet and, lumbering, roly-poly animal that it is, its plantigrade feet leave a track pattern somewhat suggestive of that of the bear's in miniature. Raccoon tracks, however, are usually paired, with the left hind foot placed beside the right forefoot as the animal walks along (see figure 96, b). This unique trail pattern makes raccoon tracks much easier to identify. Front tracks measure 1¾–3⅛ inches long and 1½–3¼ inches wide. Hind tracks are 2⅛–3⅞ inches long and 1½–2⅝ inches wide.

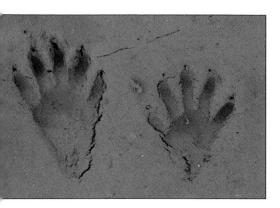

Exceptional front and hind tracks of a raccoon in central California.

Raccoon scats are by no means as readily identified as the tracks. They often have a granular appearance, tend to be even in diameter, and break easily when disturbed. In color they range from black to reddish, sometimes bleached to white. Many samples are irregular in shape, dependent on their varied diets. On the whole they might be confused with those of skunks, opossum, or bear cubs. Often scats are deposited on large limbs of trees or logs, in which case that fact is indicative. Also look for latrines of many scats at the base or in the crotches of large trees, especially those near water, or under rock overhangs in outcrops or cliffy areas. This animal is omnivorous and relishes flesh, fruit, nuts, corn and other garden crops, and carrion.

In Minnesota along the Red River, we used to identify the raccoon trees by the claw scratches on the bark, made as the animal climbed. They were usually elm trees, but occasionally oak or basswood. First we looked for a hole in the trunk, where a large limb had broken away and a cavity had been formed in the trunk. Having found a big tree with such an apparent cavity high on the trunk, we would check by looking for the scratches on the bark. In more southern woods the sycamore may be the raccoon tree, as in figure 96, d. Raccoons also utilize a variety of dens in the ground. They may use a hollow log or cavities among rocks. A most unusual instance is reported by W. H. Bergtold, who found a raccoon with a family of young housed in a magpie nest!

Although I have lived in raccoon country many years, I cannot describe the call from personal experience. Yet several writers have described this animal as quite vocal, and Seton, referring to its call as the "whicker," describes it as a "long drawn tremulous 'whoo-oo-oo-oo.'" He compares it with a querulous call he says it

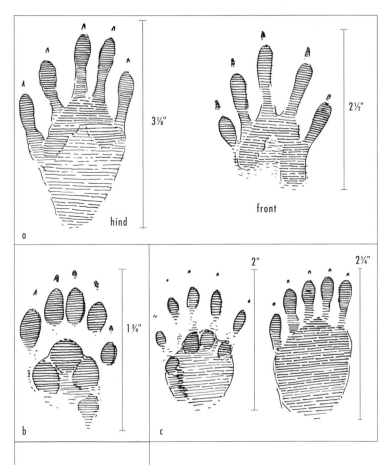

Figure 95. Tracks of raccoon and ringtail
families, drawn to scale

a. Hind and front tracks of raccoon.
b. Track of coati.
c. Track of kinkajou.
d. Track of *Bassariscus astutus*, or ringtail.

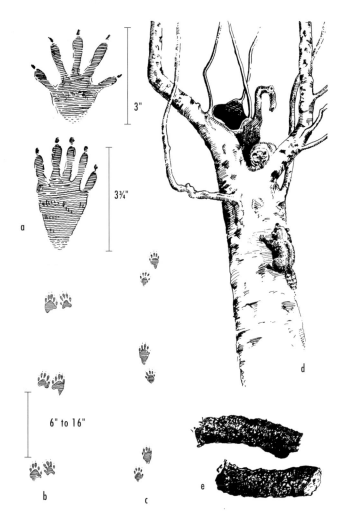

Figure 96. Raccoon

a. Typical tracks in mud, the front track above.

b. The distinctive paired track pattern, so characteristic of walking raccoons (Colorado).

c. A fast walk in moist sand (Oregon).

d. Raccoon tree, a sycamore (Arizona).

e. Rather typical droppings, with lack of taper at the ends.

may be confounded with, the call of the screech owl. Squabbling raccoons are very vocal and their shrieks and screams may send shivers down your spine.

WHITE-NOSED COATI

Related to the raccoon, and sharing its inquisitive traits, is the white-nosed coati, *Nasua narica*. Its true home is the southern tropics, but its range extends north through Mexico into the southern border of Texas as well as southern Arizona and New Mexico.

This venturesome long-tailed and long-snouted animal is a successful inhabitant of the jungle, and with encouragement becomes very friendly with people. At the headquarters of Barro Colorado Island, in the Canal Zone, some of these coatis had learned to walk on a slack wire, balancing precariously with the agile body and long tail, to seek the pieces of bread suspended near the middle of the wire. In the wild, their food habits appear to be similar to those of the raccoon. They are fond of fruits and other vegetation, and pick up any birds' eggs or young birds they may find, as well as insects or other small creatures they are able to capture.

As with the raccoon and opossum, the scats will vary in appearance in accordance with the type of food eaten.

The tracks shown in figure 97, a through c, are from captive animals, but the trail in e was that of a coati entering a papaya stand in southern Mexico. This is their normal walking pattern. When startled, coatis lope or gallop away more gracefully than raccoons.

The perfect hind track of a coati, registering the heel completely as the animal was loping in dust, in southwestern Mexico.

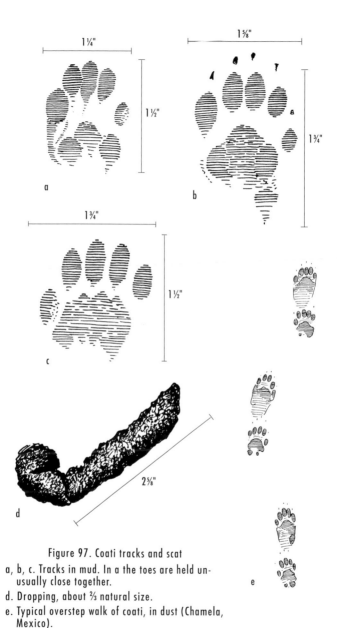

Figure 97. Coati tracks and scat
a, b, c. Tracks in mud. In a the toes are held unusually close together.
d. Dropping, about ⅔ natural size.
e. Typical overstep walk of coati, in dust (Chamela,
Mexico).

Front tracks measure 2¼–3¼ inches long and 1¼–1⅞ inches wide. Hind tracks are 2⅝–3⅝ inches long and 1⅜–2 1/16 inches wide. Coatis are tremendous climbers and, like squirrels and fishers, can turn their hind feet to face backwards. This allows them to descend trees headfirst, rather than rear end leading as with raccoons.

KINKAJOU

Kinkajou, *Potos flavus*, is only a name to most of us, suggesting that vague world we know as the tropical jungle—and perhaps suggesting the traditional "tooth and claw" so commonly associated with that lush green world. As a matter of fact, when kept as a pet the kinkajou has a genial, friendly disposition.

This animal ranges northward into tropical regions of Mexico, and thus becomes a member of the North American fauna. During a few days spent in the Panama Canal Zone I had hoped to have a glimpse of a kinkajou, possibly among the limbs of a tree, using its prehensile tail as I saw the white-faced monkeys doing. But I saw none, nor did I find the tracks.

For the tracks in figure 98 I called on a cooperative kinkajou in the National Zoological Park that allowed itself to be handled, and accommodatingly stepped in some mud. The tracks suggest those of the raccoon—not surprisingly, in view of its close relationship.

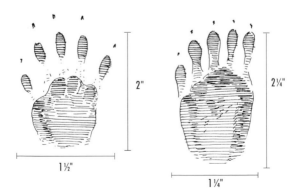

Figure 98. Front and hind tracks of captive kinkajou, in mud

Family Mustelidae:
Weasels, Otters, and Badgers

In the game, or profession, of animal tracking, it will add interest and value to your efforts to keep in mind group characteristics. The weasel family is a vigorous group of animals that have dispersed themselves into all conceivable niches of our natural environment. The otter and the mink took to the water, the sea otter developed a marine life, the fisher and marten found the northern forests, and the wolverine developed a powerful physique and took to the timberline and the subarctic. The badger, with powerful forelimbs, explores underground. And the agile little weasels, bold and persistent hunters with a great zest for life, explore everywhere.

The feet of this family have five toes, both front and hind (figure 101), though the fifth toe may not be evident in every track. There is also a tendency in this group to form a twin-print pattern when traveling in deep snow (figure 103, c). Perhaps the scats shown in figure 102 illustrate again how similar they are. There is much overlapping in size between them.

As you note the differences in the tracks of the weasel family in the field, you will be able to read in the trail something of the diverse characters and mental traits of the animals that left the footprints.

AMERICAN MARTEN

The American marten, *Martes americana,* is a dry-land counterpart of the familiar mink, inhabiting the boreal forests of the continent, and forested mountain ranges of the West. Martens are tremendous climbers and like the fisher can turn their hind feet 180 degrees, allowing them to run headfirst down trees.

The loping trail of a marten in Tahoe, California.

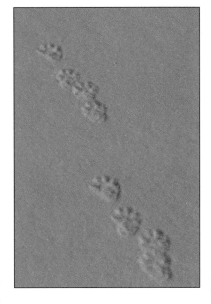

Marten tracks may be confused with the slightly smaller mink and the larger fisher. All of these, as well as the weasel, have similar but varying track patterns overlapping in size. For example, Antoon de Vos of Canada finds that the walking stride of a male fisher is 13 inches, of the female 9 inches; the similar stride of the male marten is about 9 inches, that of the female 6 inches.

Compared to the mink's, the marten's feet are larger, but there is much variation in size of footprint. The front tracks of martens measure 1⅝–2¾ inches long and 1⁵⁄₁₆–2⅝ inches wide. Hind tracks are 1½–2¾ inches long and 1³⁄₁₆–2¼ inches wide. The straddle of the marten's twin-paired trail varies from 2½ to 4½ inches. Look also for snow tunnels in dry fluffy conditions and for "body prints"—the impression of the entire marten after it has launched itself from a tree and landed in the snow. In the Sierra Mountains, I once measured such a jump from a large tree by a female marten I was following. She landed nearly 19 feet from the trunk!

In midwinter the undersurface of the marten's feet is so heavily covered with hair that the toe pads do not show. Toward the end of winter the toes appear, and in summer they are prominent.

Marten scats are confusing. They are similar in shape to those of mink and weasels and are about the size of those of the mink; they may sometimes overlap in size with those of a large weasel. One circumstance that is helpful is the fact that martens are fond of blueberries, huckleberries, mountain ash berries, and pine

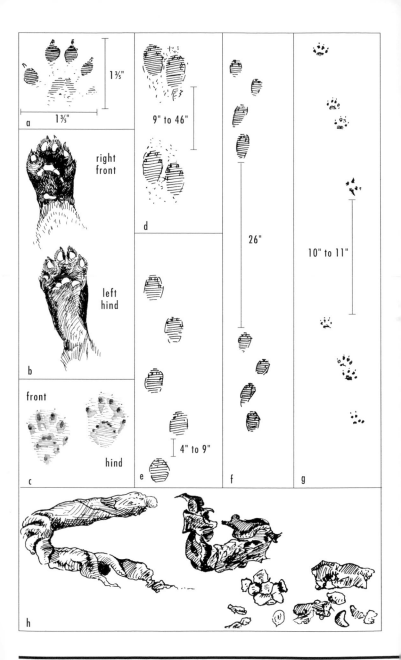

nuts, which occur less frequently in the diets of mink and weasels, and these fruit contents characterize many marten scats. In Alaska it was found that the lips of marten were stained blue in berry time. Marten scats may be found on rocks along mountain trails, as well as in the trail itself, which is unusual for fisher and may provide a clue when you are discerning scats in an area where they both occur. Often several are found together, showing the marten's tendency to deposit them where others have already been left, as many other animals do. Weasel and marten droppings may also be found together.

Marten dens are normally in a tree, in a convenient cavity, though dens in the ground have been reported. The marten will also bury surplus food, as do weasels and mink.

The marten has no prominent characteristic call, though in distress it will hiss, or growl, and sometimes "scream." In the wild it has no call that is useful as a guide to its whereabouts.

As in the case of weasels, birds such as magpies, jays, robins, and other small birds in the vicinity may assemble to scold a marten prowling in the daytime.

FISHER

The fisher, *Martes pennanti,* is a larger relative of the marten that has become very scarce in the Rockies but is doing better in the Sierras and coastal mountains into Oregon. In the Northeast, as in New England, fisher populations have rebounded astonishingly in recent years, and their signs have become common. Like marten, fishers spend part of their time hunting the canopy, as they are excellent climbers. Interestingly, it is the female fisher that most often takes to the trees, while the much heavier male

Figure 99 (opposite). Marten

a. Print on hard snow.

b. Feet in summer (Wyoming, August 1, 1928). In winter thick hair on the soles conceals the toe pads, though they begin to show again in late winter.

c. Front and hind tracks in shallow snow (Idaho).

d. A common trail in deep snow, where the hind feet register in the front tracks.

e. Walking tracks, where hind tracks register upon front tracks.

f. A gallop seen on harder snow.

g. Typical loping gait on crust, toes showing. The right front track reveals an injured foot (Teton Mountains, February 24, 1946).

h. Three samples of scats, about ⅔ natural size.

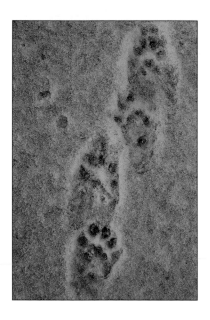

Loping tracks of a fisher in shallow snow in Peterborough, New Hampshire.

usually hunts on the ground.

The fisher is another lithe member of the weasel family, and its tracks may at times be confused with marten or otter tracks. As Antoon de Vos puts it: "The average walking stride of a large male fisher is 13 inches, that of a large male marten around 9 inches, while that of a female fisher is about 9 inches and of a female marten 6 inches." But remember that a fisher's weight is three to eight times that of a marten, and weight influences tracks and trails.

Stride and size of tracks are extremely variable, depending on snow condition and gait. Fishers may leap up to 6½ feet, though this is unusual. A fisher's most common gait, typical of all members of the weasel family, is a loping 3-by-4 pattern (see figure 100, d). In deep fluffy snow, though, as with other weasels, you will most often find fishers in the characteristic twin-track pattern, or walking (see figure 100, e).

The front tracks of fishers measure 2⅛–3⅞ inches long and 1⅞–4¼ inches wide. Hind tracks are 2–3⅛ inches long and 1½–3½

Figure 100 (opposite)
a. Front track along a riverbank (Massachusetts).
b. Front and hind left feet.
c. Walking tracks, in mud (Olympic Mountains, Washington, April 5, 1934).
d. Typical loping gait of fisher in shallow snow (New Hampshire).
e. Common track pattern in deep snow (Sequoia National Park, 1941).
f. Scat, reduced about ½ in size (British Columbia, 1934).

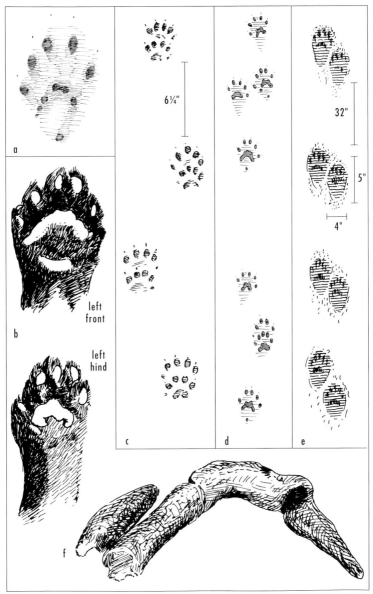

left
front

left
hind

6¼"

32"

5"

4"

a

b

c

d

e

f

Figure 100. Fisher

inches wide. As with martens, a fisher's claws are semiretractable and may or may not appear clearly in tracks.

Fishers have a habit of climbing trees, the females more often than the males, and can readily travel from tree to tree. Under such circumstances it is necessary to circle widely to pick up the trail in the snow farther on. In the Sierra Nevada, at least, the fisher has been found to burrow in the snow after mice, but apparently such undersnow travel is rare in much of its range. Like martens, fishers may descend trees head first, or launch themselves from above, leaving a characteristic body print in fluffy snow, before loping away to hunt elsewhere.

The droppings may be confused with those of the marten, though they are usually larger. The fisher is quite omnivorous and feeds on berries and nuts as well as flesh, in this respect resembling the marten. The fisher also feeds on porcupines and swallows a certain number of quills. The coyote does the same, so that the scats of both of these animals may contain porcupine quills. Coyote scats, however, are larger, except in the case of small pups.

The fisher has no outstanding call aside from the usual growls of animals that are threatened. However, fighting fishers emit varied shrieks and screams which make the skin crawl.

Their natal dens are most often in trees, but they may nap or rest in tree cavities or ground burrows.

ERMINE, LONG-TAILED WEASEL, AND LEAST WEASEL

Weasels are distributed over the whole continent and are familiar to most people. Familiar, too, should be their 2-by-2, or "paired,"

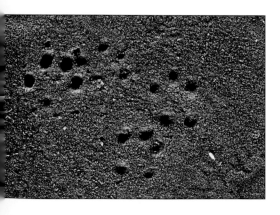

Ermine tracks along the Little Snake River in southern Wyoming.

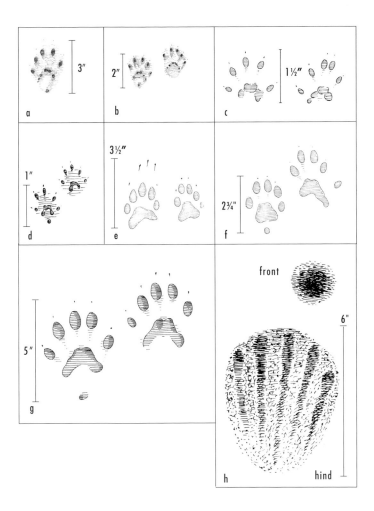

Figure 101. Tracks of Mustelidaea

a. Fisher.
b. American marten.
c. American mink.
d. Weasel, *Mustela* sp.

e. American badger.
f. Northern river otter.
g. Wolverine.
h. Sea otter.

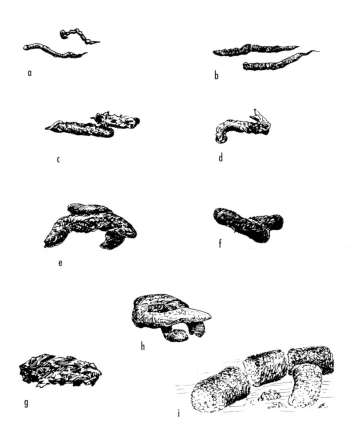

Figure 102. Droppings of the weasel family, in proportionate sizes;
the diameters given are variable

a. Ermine, *Mustela erminea;* diam. ⅛–⁵⁄₁₆ in.

b. Long-tailed weasel, *M. frenata;* diam. ³⁄₁₆–⅜ in.

c. American mink, *M. vison;* diam. ¼–⁷⁄₁₆ in.

d. American marten, *Martes americana;* diam. ³⁄₁₆–⅝ in.

e. Wolverine, *Gulo gulo;* diam. ⅜–1 in.

f. American badger, *Taxidea taxus;* diam. ⅜–¾ in.

g. Northern river otter, *Lontra canadensis;* diam. ⅜–1 in.

h. Fisher, *Martes pennanti;* diam. ³⁄₁₆–¾ in.

i. Sea otter, *Enhydra lutris;* diam. 1½ in.

track pattern in snow country. Weasels traveling in search of prey leap here and there in energetic fashion, and you will find their slender snow trails suddenly changing direction, doubling back, looping around here and there, disappearing under a half-buried log to reappear farther on. Weasel tracks are eloquent. Looked at knowingly, they reveal the character of the nosy, eager little hunter.

When loping in snow the weasel's hind feet usually register in the front tracks, nearly or completely, so that the trail appears as a line of twin prints (figure 103, c). Usually, these fall one slightly ahead of the other. Often, too, one leap is short and the next one long, producing an irregularity in spacing—alternating short and long. In deeper, drier snow the footprints of the short leap are connected by a drag mark, thus producing a very characteristic pattern (see figure 104).

One day we were walking through the winter woods in Wyoming when my companion said, "Look at the weasel!" Some hundred feet away, a white face accented with black eyes and dark nose protruded from the snow, regarding us intently. It ducked out of sight, then reappeared a few feet away. This was repeated several times, and then we saw it no more. This is characteristic, the weasel readily diving into snow as if it were water. Often the trail will disappear into a neat round hole in the snow, to reappear some distance beyond.

Although the twin-print pattern is a common one, there are many variations. See figure 103, e, for these. You are more likely to find these lope variations, groups of three and four tracks, when tracking weasels away from snows.

The accompanying sketches of tracks require interpretation. Space does not permit inclusion of tracks of the numerous species, which vary greatly in size. The three basic species are ermine, or short-tailed weasel, *Mustela erminea* (about

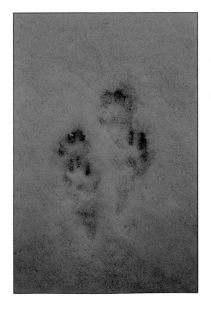

Tracks of a long-tailed weasel in snow in Pisgah State Park, New Hampshire.

15 inches long in the North to about 9 inches in the Southwest); long-tailed weasel, *M. frenata;* and least weasel, *M. nivalis.* The least weasel is about 6–9 inches long, and the long-tailed weasel may be from 12 to 20 inches long. Each group has a variety of subspecies of different sizes, and mustelids exhibit dramatic sexual dimorphism, meaning the two sexes of the same form are strikingly different in size. Add to this variations due to the character of the snow or earth, and you have a complex scale of size values. However, weasel tracks fall into a general pattern, and if one determines which weasel forms occur in a given locality, tracks of the basic species may be distinguished fairly well.

The average length of a weasel's stride may be helpful when determining species, but considered alone it cannot be used to reliably differentiate between weasels. *M. frenata* may leap 20–51 inches, while in the same locality *M. erminea* more often leaps 13–40 inches. There is tremendous overlap in stride length. Further confusion can be expected from long-tailed weasels slowly exploring niches for prey. In Alaska some least weasel tracks showed leaps up to 23 inches.

There is a more significant difference in the width of the pattern, or straddle. For the least weasel it will be about ⅝–1¼ inches in soft snow. For *M. frenata* in Wyoming it may be as much as 3 inches in width. Individual track width may be more useful. In New England, ermine tracks measure up to ¾ inch wide, and long-tailed tracks measure ¹¹⁄₁₆–1³⁄₁₆ inches wide. There is little overlap, yet be warned that the northern ermine, such as those in Alaska, are much larger animals; of course, in the Far North there are no long-tailed weasels present with which to confuse their tracks.

Weasels have five toes, but the fifth toe does not always show. In fact, it is only under exceptional circumstances that all the toes will show clearly. Their feet are also well covered with fur.

Figure 103 (opposite)

a. Tracks of long-tailed weasel, *Mustela frenata,* in wet snow. Leaps were 29–34 in.

b. Tracks of long-tailed weasel in mud (Massachusetts).

c. Trail of ermine, *M. erminea,* in snow, carrying a lemming (Hudson Bay, 1915).

d. Tracks of ermine on sand (Kodiak Island, Alaska, September 1936).

e. Trail in snow of long-tailed weasel (Wyoming, 1949).

f. Feet of ermine, about natural size (Cascade Mountains, Oregon, 1913).

g. Ermine droppings, about natural size (Wyoming).

h. Long-tailed weasel droppings, about natural size (Wyoming).

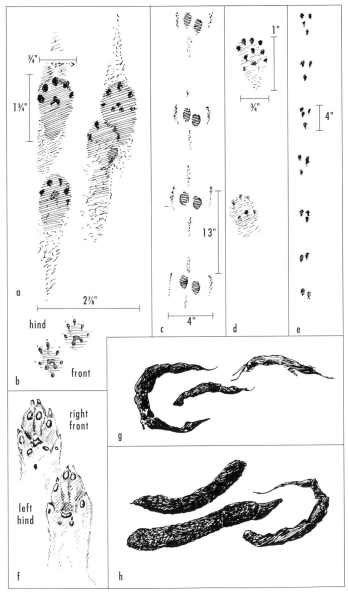

Figure 103. Weasels

The scats are long and slender, dark brown or black in color. They vary in size between subspecies but may overlap in measurements. In practice, a large collection of weasel scats may be assorted between larger and smaller forms, probably with few errors. But it would be difficult to allocate properly a single scat specimen in the intermediate size range. Scat contents may be expected to consist of rodent fur and bits of bone, more rarely feathers.

An occasional dropping may be found by following a snow trail. An accumulation will be found at a den or near a winter nest under the snow. In summer droppings are deposited along trails or roads, often on rocks or other prominent objects in or beside the trail. Such places are often used repeatedly so that three or four may be accumulated in one spot. I have found weasel droppings on or near coyote droppings.

Dens may be in the ground (in a mole or pocket gopher hole), under a barn, in a pile of stored hay, under rocks, or in similar safe retreats. On several occasions I have found a winter nest of a meadow vole appropriated by a weasel and lined with the fur of its victim. A pile of weasel droppings will sit at its door.

Weasels store food for future use. A heap of dead mice may be uncovered in a shed of baled hay, where a weasel had found a safe retreat, or stored mice may be found under a stump or in a burrow. On one occasion, 27 mice were found next to an ermine nest in the basement of a New Hampshire house. On another occasion, 17 shrew mole carcasses were neatly piled next to a nest of a long-tailed weasel built in the engine compartment of an abandoned car in Washington State. Figure 103, c, shows ermine tracks with drag marks in the snow on either side. This was sketched in the Hudson Bay area, where I followed the trail nearly half a mile to find that the animal had stuffed a lemming down in the snow beside a bush.

Weasels are inquisitive animals, and often, when you have seen one disappear in a rock pile or a hollow tree, he will reappear to have another look if you make a mousy squeak. When greatly disturbed, a weasel will also emit a characteristic odor, to some as disagreeable as that of the skunk.

In summer you may hear a noisy outcry among birds. Note in which direction the birds are looking, and move cautiously. You may see the creature they are mobbing, possibly an owl or a crow, and again it may be a weasel. One should always listen to the warnings of birds.

It is on occasion helpful to the naturalist to know the killing technique of a carnivore. The weasel apparently prefers to seize its prey at the back of the skull, sometimes the neck or throat. Small tooth marks there would suggest the work of a weasel.

Figure 104.
Short leaps of an ermine connected by drag marks in deep snow, alternating with long leaps entirely clear of the snow. Weasels also dive into the snow, to emerge some distance away.

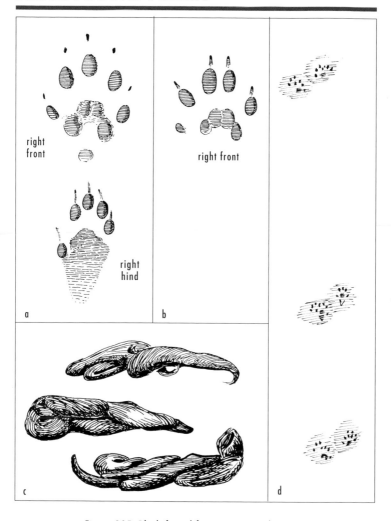

right
front

right front

right
hind

a

b

c

d

Figure 105. Black-footed ferret sign (Douglas, Wyoming)
a. Front and hind tracks (about ⅔ natural size).
b. Front track of mink for comparison (about ⅔ natural size).
c. Black-footed ferret scats (natural size).
d. Typical paired lope of ferret in light snow (Badlands National Park).

This may be the rarest North American mammal. *M. nigripes* science calls it, denoting its close kinship within the weasel group. At one time it was found generally throughout the prairie dog country of the West, though apparently it was never really abundant. With the drastic poisoning of prairie dogs, on which the ferret largely depended for food, it was reduced to near extinction. This endangered species survives mainly in South Dakota. Comparatively few people now living have seen this animal alive. Through the kindness of Mr. Warren Garst, who had the privilege of studying a few captive animals in eastern Wyoming, tracks and droppings were made available to me. (These animals were later released in a national park.)

Figure 105, a, illustrates the tracks in natural size. Note the close resemblance to the mink track, shown for comparison in b. Probably it would be very difficult to distinguish them in the field. The droppings, too, are similar to those of mink, and have the same type segmentation when composed largely of hair.

However, if you come upon tracks like these in a prairie dog town in Badlands National Park, for example, or in prairie dog towns elsewhere in the West, they will probably be those of the black-footed ferret.

AMERICAN MINK

Like the weasel, the American mink, *Mustela vison*, occurs over most of North America, but it is more restricted to forest cover and water. The mink and the weasel are fashioned after a similar pattern, and their tracks are similar in form. Mink tracks in deep snow are generally in the familiar double-print pattern, a larger edition of those of the weasel. This pattern is made by the hind feet almost registering in the front track (figure 106, c). The mink, too, will go down under the snow on exploratory dives. When traveling along, the mink will occasionally push itself forward in the snow, leaving a trough, and it will sometimes coast down a slope like an otter. It is obvious that these dives into snow and the coasting downhill on snowy surfaces reveal in the mink a degree of the exuberant playfulness we find in the otter.

Along a river in winter you may find a smooth round hole down through the snow and through an air hole in the ice, where the mink has been foraging under water. The hole may be more or less muddy, for the animal has been on the stream bottom, and there may be fragments of frogs or other food nearby.

As in the case of the weasels, the male mink is larger than the female, and their tracks differ correspondingly in size. Trappers often attempt to distinguish tracks of male mink from those of fe-

Loping mink tracks along the Connecticut River in Massachusetts. Of the two tracks at top, the smaller, above, is the hind track; the one just below is the front.

males, and to some extent this may be possible, with experience. Condition of snow and age of the animal as well as sex affect the size of tracks. Front tracks of the mink measure 1⅛–1⅞ inches long and ⅞–1¼ inches wide, including the claws when present. Hind tracks range from ¹³⁄₁₆ to 1¼ inches long and ¹⁵⁄₁₆–1⅝ inches wide. The width of the track pattern, or straddle, varies from 2 inches in mud to 3¼ inches in snow. Running gaits in mud and on firm snow are illustrated in figure 106, d and e. Note similarity in mink gait, d, to skunk gait in figure 118, d.

Mink scats are of course somewhat larger than those of the weasels, though an occasional small one will fall within the weasel size. When consisting of fur, as the dark one in figure 106, f, they are irregularly segmented, or folded, and are blackish in color. The light-colored one in this figure is characteristic of those consisting of feathers. If the contents are fish remains or similar food, the scat is rough and usually black and glistening. If composed of crayfish remains, the scats are reddish.

Figure 106 (opposite)

a. Typical mink tracks in mud.
b. Mink feet. Note extra round pad at posterior of front foot.
c. The common mink trail in deep snow, twin prints made by the hind feet registering upon the front tracks, or nearly so.
d. The common loping gait of the mink.
e. A high-speed bound.
f. Mink scats, about ⅔ natural size.

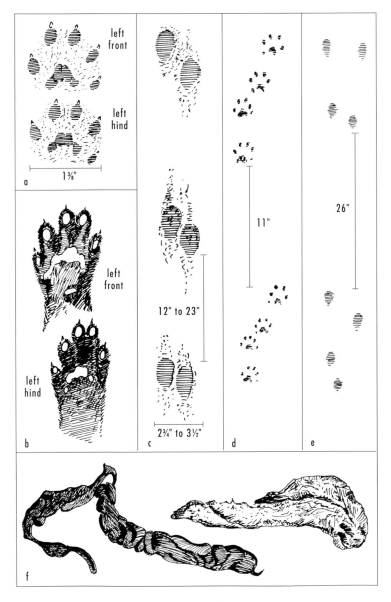

left
front

left
hind

1 ³⁄₈"

a

left
front

left
hind

b

12" to 23"

2 ³⁄₄" to 3 ½"

c

11"

d

26"

e

f

Figure 106. Mink

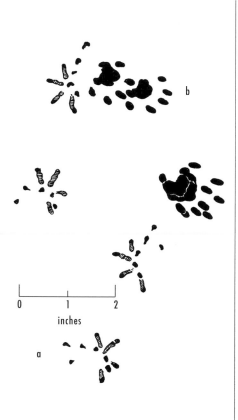

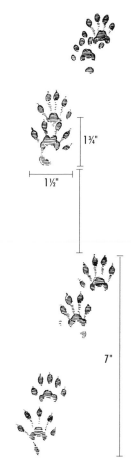

Figure 107. Tracks of mink and toad traveling in nearly opposite directions, in soft mud

a. Toad tracks, walking gait showing only the toe tips of the hind feet superimposed on the 4-toed front-foot tracks.

b. The mink also had stepped in or nearly in the front-foot tracks with the hind feet. Here again is illustrated the making of the 2-by-2 pattern, the 4 feet making what appears to be 2 tracks when details are lost in snow.

Figure 108

Typical loping pattern of mink in mud (Wyoming). This is similar to the trail shown in figure 106, d, but note that the individual tracks tend to segregate into pairs, approaching the 2-by-2 pattern shown in figure 106, c.

Mink dens may be dug into banks, in which case they will be approximately 4 inches in diameter. The mink may also use muskrat burrows, holes in logs or stumps, or other ready-made shelters. The nest is made of leaves, in some instances at least, and is about a foot in diameter. It may be lined with feathers, when these are available.

A beaver pond often gives us a story of adaptation. The beaver has furnished the engineering skill to create a pond. The muskrats take advantage of this and make their home in the same pond, often with shelters within the outside structure of the beaver house itself. Then along comes the mink, which is fond of muskrat meat, and preys upon some of these animals. On several occasions I have found on the beaver house a collection of mink droppings consisting of muskrat fur and bone.

Like the weasel and the skunk, the mink produces a strong scent, but not as disagreeable as that of its two relatives. As boys, some of us used to poke a long switch into a suspected mink den. If we then caught the scent given off when the mink is disturbed, we knew that the animal was at home.

TAYRA

The tayra of the American tropics, *Tayra barbara*, reminds one of the marten and fisher of the North, and the track, shown in figure 109, is much like that of the fisher. They need not be confused, however, for the fisher is confined to the snow country of part of

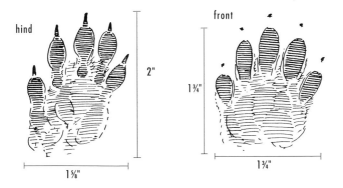

Figure 109. Hind and front tracks of a pet tayra
(Panama Canal Zone, September 1952)

the Rocky Mountains and the northern forests, while the tayra lives in South and Central America and comes north only as far as parts of Mexico.

I had not found the tayra or its tracks in the jungle of the Canal Zone, but Lieutenant H. H. J. Cochran of the Air Corps, who was stationed there, had a pet tayra, and his wife saw to it that I got the tracks. Mrs. Cochran works with ceramics. After she had mixed some clay in two dishes, they maneuvered the unruly, lively pet until it pressed its feet in the clay, front foot in one dish, hind foot in the other. So, through the enthusiastic help of two animal lovers, I was able to get at least a couple of samples of tayra tracks.

This animal, like so many of the weasel tribe, is inquisitive, full of vitality, and certainly becomes a fascinating pet.

WOLVERINE

The most powerful and picturesque character of the weasel tribe is the wolverine, *Gulo gulo,* whose home is the boreal and Arctic regions of the Northern Hemisphere, and some of the western mountain regions of the United States.

Few have seen this elusive animal of the wilderness, which has now become exceedingly scarce south of Canada. Therefore its tracks assume all the greater importance. As shown in figure 110, perfect tracks reveal the five toes characteristic of the weasel tribe, though the small toe does not always show. There are many track patterns: walking or trotting, as in b; a little faster, c; loping, d; galloping, e. It is interesting to note some similarities with other members of the weasel family. Observe the "mink" pattern in c; compare with the similar double-print pattern in the mink, weasel, marten, and fisher. Again, compare the loping pattern, d, with the similar pattern of the skunk, mink, marten, and fisher — even to some extent similar patterns of the red fox and wolf. In f, the tracks show that the left hind foot is injured and not being

Figure 110 (opposite)
a. Typical tracks, which show all 5 toes.

b. Walking or trotting gait.

c. A 2-by-2 lope, common in deep snow.

d. An easy lope.

e. A high-speed gallop.

f. Track of wolverine with crippled left hind foot. (Note that there are only 3 footprints, signifying that it must have been holding up the injured foot.)

g. Typical scat.

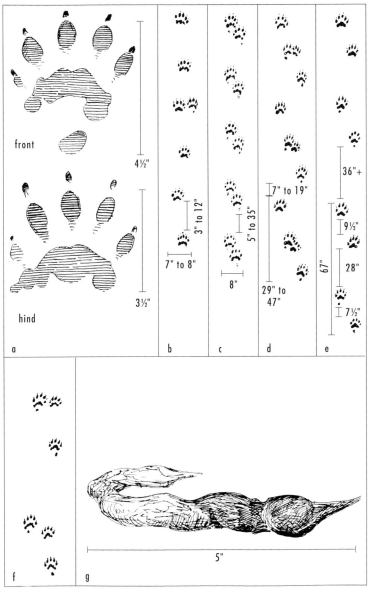

Figure 110. Wolverine tracks and scat

The front track of a wolverine near Valdez, Alaska.

used. The size of the tracks varies, of course, with the size of the animal and condition of snow. The front track may vary in length from 3⅝ to 6¼ inches and in width from 3½ to 5¼ inches; in deep snow there will only be a series of deep holes, as in figure 111.

Wolverine tracks may be confused with those of the wolf at first glance, but the wolf has four toes, the wolverine five, and the heel pads are different shapes. The wolverine trail tends to wind around, the wolverine nosing about in an inquisitive manner, alert for any tidbit of game or carrion.

Wolverine scats have the usual elongated weasel shape, but are larger than those of the other Mustelidae, except for the distinctive scats of the sea otter.

The wolverine has been referred to as "skunk bear." In fact, in shape it does resemble a diminutive bear, and it has two pale brown stripes on the sides that are somewhat skunklike, reaching to a bushy tail. Moreover, it has well-developed anal glands that produce a disagreeable odor.

The only sounds I have heard from wolverines were growling and snarling, among a group of captive ones in a zoo.

The wolverine has a notorious personality that has been a popular subject in journalistic writing. There is no question about its ferocity and strength. I wonder if there is another inhabitant of the northern wilderness that so excites the imagination. One time I came on a wolverine trail in an early winter snowfall in central Alaska. So eager was I for "wolverine lore" that I laboriously tracked the animal a long distance to see what it had been up to.

Merely seeing those tracks in the snow made it a red-letter day. Its keen nose had discovered something under the snow, which proved to be a raven wing. At another place it had uncovered some caribou bones. Thus there unrolled a record of seeking and finding animal remnants that evidently mean so much in the economy of the wolverine.

At another time I shot a mountain sheep ram for a museum specimen. It was a bitterly cold winter day, and dusk was falling. There was not time to completely skin the animal and get back to my tent. I had seen wolverine tracks on that mountainside, and I knew what could happen to my specimen if I left it there.

Figure 111. Wolverine running in deep snow

So I made a bargain with the wolverine. I didn't want him to spoil the head of the sheep. In great haste in my race with darkness, I partly skinned the ram back from the rear, laid the loose skin back over the head to protect it, and left exposed for the wolverine's feast the hind quarters and the belly, with its choice internal organs.

Next morning I went back. There were the wolverine tracks all around the carcass. Great chunks of the best meat had been taken out, but my specimen, the skin and head, was untouched.

AMERICAN BADGER

This weasel became burly and strong, spurned the trees and the water, and decided to gain his living underground. The American badger, *Taxidea taxus,* is short-legged, flat to the ground, with powerful front feet and claws for digging. Unlike the nocturnal English badger, who lives in the woods, the American badger is characteristic of plains country, from Mexico north into southwestern Canada, from the Pacific Coast eastward to the Middle West, including parts of Ohio, Indiana, Illinois, Kansas, Oklahoma, and Texas. To some extent it also ventures into high mountain country and into woodland.

The badger hole, the most prominent sign, is a feature of the western landscape, known to the early riders of the range as a hazard for their saddle horses, since stepping into one unknowingly could cause a broken leg for the horse and a bad fall for the rider. Abandoned holes also furnish nesting places for burrowing owls and refuge for other animals.

The burrows are conspicuous, with entrances from about 8 inches to a foot in diameter, elliptical in shape, as one would expect from such a flattened digging dynamo. A large mound of earth is thrown out, for the tunnel is big and a great deal of earth is moved. There may be many holes in a given area, since the badger seeks food by digging for one ground squirrel or other rodent after another. In prairie dog towns, numerous burrows are enlarged by the badger, who is after these rodents.

Figure 112 (opposite)
a. Tracks in dust (Jackson Hole, Wyoming, 1931).
b. Trotting trail in dust (Jackson Hole, 1931).
c. Walking trail on hard snowdrift, fragmentary type often found (Jackson Hole, December 22, 1927).
d. Badger trail in deep snow.
e. Badger scat, about ⅔ natural size.

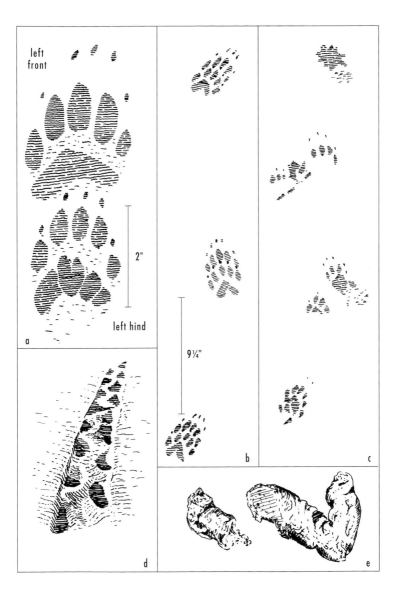

left
front

2"

left hind

a

9¼"

b

c

d

e

Figure 112. Badger tracks and scat

The walking trail of an American badger in northwestern Colorado.

Badger tracks, shown herewith, are extremely toed-in, and the long claws of the front feet generally leave marks. Figure 112, c, illustrates a badger trail of a fragmentary nature on hard snowdrift. Walking strides vary from 5½ to 9¾ inches, and the straddle from about 5 to 7 inches, though the width of the badger trail appears greater in loose snow. When trotting, strides may reach 15 inches between tracks and the trail width will narrow significantly. The front track is 2⅞–3⅞ inches long (including nails) and 1 ⁹⁄₁₆–2⅝ inches wide. The hind track is significantly smaller at 1⅞–2¾ inches long and 1⅜–2 inches wide. Sometimes the hind-foot track is behind that of the front foot, as in figure 112, b and c, but at other times it is in front. Note also that the hind track often only registers four of its five toes.

I have not found the scats prominent, at least to casual observation. Nor have I discovered any outstanding characteristic that would distinguish them readily from those of other carnivores of equal size. Again, one must be alert for accompanying signs that give additional clues, and keep in mind what animals are known to occur in a given place.

Whenever you meet a badger you will be impressed by its definite character. The only sound I have heard it make is a growl, or a hissing sound, as it faced me, daring me to come closer. At other times, when it poked its head out of a hole to have a look at me, its striped face has had an almost clownish look.

Once a badger started to dig into the ground to escape. I seized it by the hind legs and tried to pull it out of the hole as the hind quarters were disappearing, just to have another look at it and to see what it would do. But the badger held fast. I felt as if I were trying to pull out a big plant by the roots. In a few moments I noticed the muzzle coming out, doubling back under the belly, reaching for my hands. I promptly let go and watched it disappear into the ground!

NORTHERN RIVER OTTER

Though not commonly observed in the wild, the northern river otter, *Lontra canadensis,* is a fairly familiar animal to most people. Its original range covered most of North America. It is one of the larger of the weasel tribe, its recorded weights running from 5 to nearly 30 pounds.

The otter is agile, as fluid in its movements as the water that is its favorite element. Yet on the land it is not as light on its feet as the weasel or fisher and seems almost to plow through the snow. This is revealed by its tracks, which sometimes appear in a snowy trough. Characteristic, too, is the long mark in the snow where the otter has slid. Coasting is enjoyed occasionally by the mink, but the sport is developed to the extreme by the otter (figure 114).

One wintry day in southern Hudson Bay territory I was snowshoeing up a small stream when I spied a movement on the snowy streambank ahead. I realized that it was an otter, and the next moment it slid down the bank. Another one appeared, clambered up the bank, and slid down. A third appeared from a hole in the ice, and for several minutes I watched these frolicsome animals, climbing, sliding, climbing, sliding, over and over again

The typical pattern of a loping river otter, along a riverbank in southern Alaska.

—until they all disappeared under the ice. Their playtime was over, and they went on their way beneath the ice, as so often they do.

When coasting thus the front feet are held back along the sides, and the hind feet trail out behind, in a streamlined arrangement. These marks in the snow are sure evidence of the otter's presence. Some of these slides may be as much as 25 feet long, or longer. The otter will also slide on the level, giving himself a forward push as he travels along. Furthermore, both the otter and the mink may dive into loose snow when closely pursued, to come up some distance beyond. Severinghouse and Tanck (1948) recorded level slides on ice as long as 20–25 feet, when the otter's speed was 15–18 miles per hour. The slides were shorter on well-packed snow. The late Francis H. Allen kindly furnished the following incident:

> On March 7, 1937, I found otter tracks on the snow-covered ice at the edge of a pond in Cohasset, Massachusetts. The ice extended only a little way out into the pond, and there was no snow on the land, so that the only tracks I could see were those in the thin coating of snow on the ice. The ice was too thin to bear me, but, standing on the shore, I estimated that the otter, after a run ending a few feet from the shore, slid for about two and a half feet, then ran for about nine feet, then slid about twenty feet, then, after a run of six feet or so, slid two or three feet to the end of the snow, with open water only a few feet ahead. What was particularly astonishing to me was the length of the twenty-foot slide after so short a run. The final spring before the long slide seemed to be from all four feet at once and was evidently very powerful.

Like the mink, too, the otter will spend much time under the ice of a stream, evidently finding air space near the shore, where the ice often slopes down from a previous water level. There are usually air holes in the ice through which mink or otter can go in or out. Such holes, leading down through the covering snow and through an opening in the ice, have been described for the mink.

Figure 113 (opposite)

a. Tracks in wet sand (Wyoming, 1937). The front track is above, the hind below.

b. Tracks in deep snow trough.

c. Tracks of otter running in soft mud (Wyoming, 1936).

d. The typical loping gait of a traveling otter (Wyoming, 1936).

e. Slowing from a lope to a walk, in soft sand (Yellowstone Lake, 1937).

f. Otter scat, about ⅔ natural size.

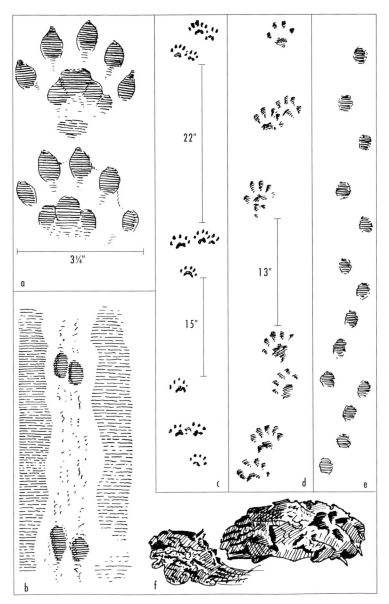

Figure 113. River otter tracks and scat

Those used by otters are of course correspondingly larger; both animals create and maintain their own holes through the winter.

The "slide," as the otter travels in the snow, may be a foot or more wide and is easily distinguished from the one made by mink. However, it should be remembered that a beaver will also come out of the water into the snow and will make a wallowing trough. But a close study should reveal some sign of the beaver's large webbed hind feet. In the distance I have seen such a beaver mark on a snowy bank and mistaken it for that of an otter.

The otter track is similar to that of the mink, on a larger scale, as shown as figure 113, a. In the hind track especially the inner toe is often conspicuously out to one side, which helps to distinguish their tracks from those of fishers. One should also note that an otter's hind feet are larger than its front feet, as opposed to the fisher, where the front feet are larger. Therefore, determining whether front tracks are larger than hind tracks or vice versa will help to distinguish between these two species. Front tracks in the otter measure 2⅛–3¼ inches long and 1⅞–3 inches wide. Hind tracks are 2⅛–4 inches long and 2⅛–3¾ inches wide.

On a firm surface the web does not leave a noticeable mark, but appears particularly in soft mud. Very often the gait is in the 2-by-2 pattern of the mink, either in the wallowing trail, shown in figure 113, b, or on firmer surfaces. Various other gaits are shown in c, d, and e and are more common in shallower substrates like sand and mud.

In addition to the slides, which may be in snow or wet mud, the

Figure 114. The otter slide

otter also leaves indications in "rolling" places. The otter loves to roll, whether in the water or on land. Evidence of such activity on land is revealed by the disturbed vegetation, and is thought to help distribute water-repelling oils throughout its coat as well as to communicate as a scent-marking behavior.

David B. Cook published a significant experience that took place in December 1938, in New York State, when he was walking up Kinderhook Creek: "Glancing upstream, I noticed a dark object floating down in midstream. As it came nearer, it resolved itself into an otter. I watched the animal come down through two riffles, with head well up, body and tail held stiff and hind legs wide apart, very obviously enjoying a free ride. It skillfully avoided boulders and kept itself in the swiftest current."

There is another sign that I have not seen, but which is described by Grinnell, Dixon, and Linsdale (1937). To quote them:

> River otters have a unique way of twisting up tufts of grass to mark selected points where scent from their anal glands is regularly deposited. A. H. Luscomb, who has long been acquainted with the river otters in the Suisun Bay region, says that he knows of several such rolling places and "sign heaps" that have been visited regularly by almost every otter passing along a certain slough during a period of fourteen years. It is thus likely that otters, like beavers, maintain certain signposts and that these stations are visited by any adult otter that passes through the neighborhood.

Otter droppings may be found at these signposts, or on logs or rocks adjacent to or extending out into the water. The boat concessionaire at Yellowstone Lake sometimes has been troubled by the otters thus defiling the rowboats during the night. Rolls and droppings are particularly associated with small peninsulas jutting out into water sources and with narrow stretches of land between two bodies of water.

The droppings are likely to be irregular in form, sometimes merely a flattened mass of fish bones and other undigested matter. Often they consist of short pieces. The color varies greatly with the type of food eaten: black to silver when eating fish, red to green when eating crabs or crayfish.

The otter den may be merely a convenient resting place under roots or other suitable shelter. The permanent den is dug into a bank, with both underwater and outside entrances. They have been found also in a hollow log; Audubon and Bachman reported them in cypress roots, one in a hollow tree that stood at the edge of the water, with an entrance beneath. The nest consists of sticks, leaves, and grass.

Trails lead off from slides, or between two bodies of water, or to a "rolling place." Such a rolling place in tules is described as a flattened-down area some 5 or 6 feet in diameter (Grinnell et al.) littered with small piles of fish scales, the remains of degraded otter scats.

Here, then, we have a member of the weasel family specialized for aquatic life; but more significantly, specializing also in the enjoyment of life in play—whether it be rolling in the grass, rolling, diving, and gliding in the water, riding on the surface of a swift-flowing stream, or coasting on mud or snow. Surely here the superabundant energy of the weasel tribe has been directed to aesthetics of a sort.

SEA OTTER

Sea otters, *Enhydra lutris,* formerly ranged along the Pacific Coast from California to the Aleutians and into the Bering Sea, as well as on the Siberian side. Today on the American side they occur along the coast of southern California south of Half Moon Bay, along the southern Alaskan coast, and throughout the Aleutian Islands.

Detecting the presence of sea otters usually means seeing the animals themselves, rather than their tracks, for these are the most aquatic members of the weasel family. They eat and sleep and breed in the water, and need not often come ashore. A male otter may be as much as 5 feet long and weigh 80 pounds or more. The hind feet are developed into seal-like flippers.

There is one sea otter sign that may come to your attention. In the Aleutian Islands I had often heard a rapid tapping sound coming across the water from where the sea otters were feeding. This was puzzling. Months later, two friends and I were watching the sea otters on the California coast and there found the solution. We heard the same sound repeatedly, and with binoculars saw what was happening. A sea otter would dive, then come up with a clam or mussel and a small rock. The otter placed the rock on its stomach as it floated on its back, then, holding the clam with both hands, it pounded the clam against the rock to break the shell! If you hear the tapping sound, therefore, coming from feeding sea otters, train your field glasses on them and you may witness an amazing performance—an animal deliberately picking up a tool, in this case a rock, on which to break open clamshells.

Sea otters prefer the kelp beds, which are present in the Aleutian area in summer only. They do come ashore at times, preferably on rocky beaches, and there can be found the unmistakable droppings, consisting of bits of sea urchin tests, pieces of shells, remains of crustaceans—altogether a compact mass of broken

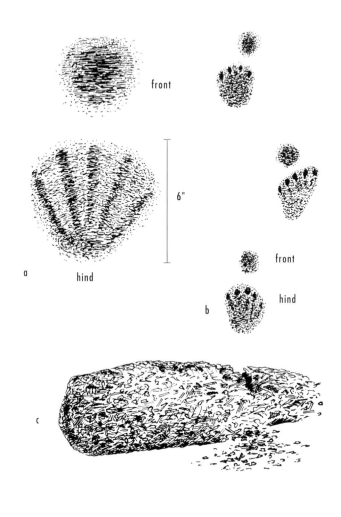

front

hind

6"

a

front

hind

b

c

Figure 115. Sea otter sign

a. Tracks in sand, showing the round front track and the larger track of the webbed hind foot.

b. Sea otter trail pattern on sand.

c. Sea otter scat, about ⅔ natural size.

bits of hard material. Sometimes you may find only the scattered grains of this material on the rocks, when the scats have crumbled. Also note that gulls will regurgitate the same kind of material. Crumbled gull casts contain but a tiny portion of the volume of sea otter droppings. Where only fragmentary remains of such material are found, there must be some doubt about identification.

Tracks are rare indeed. Once, on Ogliuga Island, a sea otter had been on a sandy beach, where footprints could be seen, as shown in figure 115.

It is often confusing to pick out a sea otter in a kelp bed, for the sea otter's head protruding from the water may be mistaken for one of the large bulbs or "floats" of the kelp.

This animal of the sea has the agility and playfulness of the river otter to a considerable degree. While lying on its back nibbling at a treasured clam or mussel, it may suddenly swirl in the water, with a combined dive and spin, then come back belly up and continue eating. Such maneuvers seem to be done just for the fun of it.

A baby will often lie curled up to doze on the stomach of the buoyant mother while she rests on the water. When she wants to go off in a hurry, she puts an arm around the baby and swims off smoothly and powerfully, apparently with great ease.

Family Mephitidae: Skunks

MOST PEOPLE RECOGNIZE the black-and-white patterns that characterize the skunks, and have learned to make way for these bold creatures without disturbing them. For the skunks are armed with well-developed scent glands near the anus that can shoot a vile and toxic substance with incredible accuracy or spray a cover of mist with which potential threats will be engulfed. Skunk spray

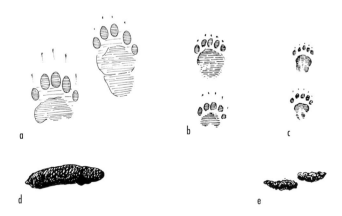

Figure 116. Skunk tracks and scats

a. White-backed hog-nosed skunk.
b. Striped skunk.
c. Western spotted skunk.
d. Striped skunk scat, *Mephitis mephitis;* diam. ⅜–⅞ in.
e. Western spotted skunk scat, *Spilogale* sp.; diam. ¼–¾ in.

may cause temporary blindness and nausea, as well as ruin any clothes you are wearing.

Skunks were once classified as members of the weasel clan, but they have recently been organized into their own family.

EASTERN AND WESTERN SPOTTED SKUNKS

The eastern spotted skunk, *Spilogale putorius*, and the western species, *S. gracilis*, are attractive little animals, more agile than the striped and other larger skunks; in fact, they are active enough to climb trees to some extent and to hunt more live prey than other skunks. They share with the larger skunks the repelling odor, and are found in varied scattered habitats from the eastern and southern states through the Southwest and Mexico to the rain forests of the Northwest. Unfortunately, the little-studied spotted skunks appear to be in decline over much of their range, making finding their tracks and signs that much more rewarding.

The much smaller tracks of this animal follow a pattern quite different from that of the larger skunks, as shown in the illustrations. Its walking gait produces a "puttering" style of track pattern, as in figure 117, e. Individual tracks can be distinguished from larger skunk species by the distinctive split heel—meaning there are two pads at the posterior edge of hind tracks rather than one. Front tracks measure 1–1 ⅝ inches long (with claws) and ¾–1 1/16 inches wide. Hind tracks are ⅞–1 ⅜ inches long and ¾–1 inch wide.

The droppings are small and irregular in shape, about ¼ inch in diameter.

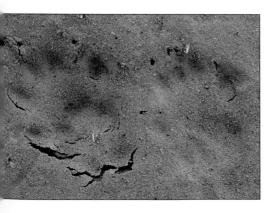

Look carefully to see the complete front and hind tracks of a western spotted skunk lightly moving along the Rio Grande in southern Texas.

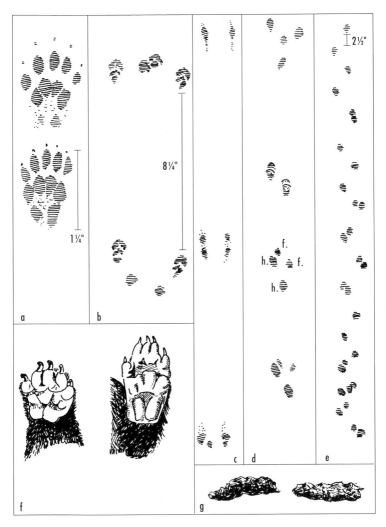

Figure 117. Spotted skunk

a. Tracks in mud (Olympic Mountains, Washington).

b, c, d. Various bounding and running track patterns.

e. Walking pattern.

f. Feet of spotted skunk; front above (Olympic Mountains).

g. Droppings, about ⅔ natural size; diam. ¼ in.

The dens are in all kinds of places—in burrows, among rocks, or under buildings. To my regret, I once disturbed a spotted skunk that had come into my cabin in the woods and sought a dark corner under the bed as a refuge!

STRIPED, HOODED, AND WHITE-BACKED HOG-NOSED SKUNKS

Under this heading are included for convenience the striped skunk, *Mephitis mephitis*, found over so much of North America; the hooded skunk, *M. macroura*, found in southern Arizona and Mexico; and the larger white-backed hog-nosed skunk, *Conepatus leuconotus*, found in Texas, the Southwest, and south into Mexico.

Their well-known defensive weapon of odorous spray from the anal glands has given these animals a poor reputation. Actually, skunks are interesting and need not cause any difficulty if approached quietly. Consider the significance of the facts. The

Front and hind tracks of a striped skunk in southernmost Wyoming along the Little Snake River.

A perfect front track of a hog-nosed skunk in northern Mexico.

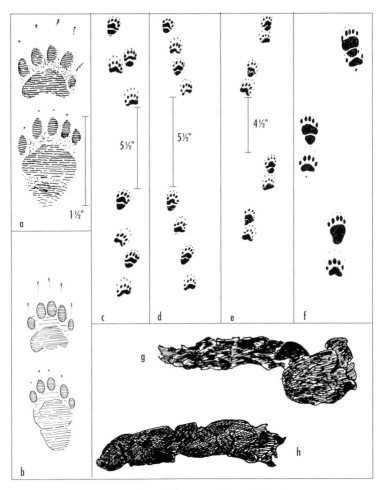

Figure 118. Tracks and droppings of striped skunks

a. Tracks of *Mephitis mephitis* in mud; the front track sits above.

b. Tracks of a white-backed hog-nosed skunk in dust; the front foot sits above (Chamela, Mexico).

c, d, e. Various loping track patterns of the striped skunk; note variations in placement of hind feet, which can be recognized by the heel.

f. Overstep walk of the striped skunk, typical walking pattern of both striped and hog-nosed skunks.

g. Scat of striped skunk.   h. Scat of hog-nosed skunk.

skunk chose to find its living by grubbing for insects, catching frogs, and eating what eggs or carrion it could find. All the weasels have the anal glands, but the skunk specialized in producing a vile liquid spray for defense. As a result of these developments and habits the skunk did not need agility. Like all sedentary creatures, including some people, it grew paunchy and stodgy. This whole history is revealed in the skunk trail. See those short steps and close pattern illustrated in figure 118.

Tracks of the hog-nosed skunk are easily distinguished from the tracks of the striped and hooded skunks by their broader palms, larger size, and square posterior edge of their hind tracks (see figure 118). The trail patterns of all three species are similar to those of bears, but in miniature. They use the same walking gaits, with hind tracks often registering beyond the front tracks (figure 118, f) and boxy loping gaits illustrated in c. Varying loping patterns are pictured in d and e.

Tracks compared (with claws):

Striped skunk—front, 1⅝–2 1⁄16 in. long, 1–1 3⁄16 in. wide; hind, 1 5⁄16–2 in. long, 15⁄16–1 3⁄16 in. wide

Hog-nosed skunk—front, 1⅞–2¾ in. long, 1⅜–1⅝ in. wide; hind, 1⅝–2½ in. long, 1⅛–1⅝ in. wide

In snow country, from New England to the Pacific, the skunk goes into the long winter sleep. But occasionally during warm spells, especially in late winter, you may find the trail of a wandering skunk winding over the snow, ending up in some den or hollow log, or under a building. Why had it shifted to another resting place? And did it later go back to the original winter nest? Does it have in its memory a series of convenient refuges? Surely it must doze a good while longer. At any rate, attention to this short-gaited line of tracks reveals interesting actions of a newly awakened sleeper.

All skunks root after insects or grubs, but this habit is more pronounced in the hog-nosed skunk, which has developed the snout for that purpose. Therefore, the diggings or extensive rooted-up places are a good indication of the presence of this animal, which in parts of New Mexico and Texas is known as the "rooter skunk." The striped skunk also digs after insects and grubs, but its work appears more as little pits. Such sign is admittedly difficult to distinguish from other disturbances.

Here again we have an auditory sign. When a skunk is under a floor, or in a similar refuge, and is disturbed or otherwise is in a nervous state, you may hear a thumping sound, made by a foot. For some time we had one under our house and in the stillness of the night we would hear these loud thumps beneath the floor. We marveled how such a loud sound could be made by a skunk. (We finally managed to plug the entrances when we were sure the skunk was away, and no doubt it has found a more remote thumping place.)

Skunks seek almost any cavity for a den—an abandoned ground burrow, hollow log, or rock crevice (even under a building!)—where they make grass nests, such nest remnants by the den sometimes revealing that it is occupied. In the case of a long-haired animal like the skunk, look also for traces of hair about the burrow entrance. Occasionally a faint smell is an indication, though normally the violent skunk odor is not emitted except when the animal is disturbed.

FAMILY FELIDAE: CATS

CAT AND DOG tracks show considerable resemblance, but there are significant differences if the tracks show details. The cats normally keep their claws retracted, hence under usual circumstances claw marks do not show in their tracks. In the cat tracks, too, the palm pads appear relatively larger, and the toes tend to be arranged somewhat as a curved row in front of the large pad, more so than in tracks of the dog family.

The droppings of canines and felines are difficult to distinguish, for their diets can be similar. Members of the cat family may scrape together dirt or rubbish to cover their dung, or create a scrape into which to place scats, and these scratch marks are useful for identifying the scat or associated tracks. However, most droppings of wild felines are left uncovered. It should be noted that coyotes and others of the dog family may also scratch with the hind feet after voiding dung, but they do not deliberately cover it.

Cats often urinate in unique fashion as well. When scent marking, felines rarely lift a leg or squat, but rather shoot urine directly backwards at some object such as a live tree, hanging vegetation, or a rotting stump. Look for their trails to meander past such objects and for a hind foot to fall unusually far from the line of travel. This is an obvious indicator that urine was sprayed.

DOMESTIC CAT

The feral cat, referred to as *Felis catus,* comes in a variety of breeds, sizes, and colors, but so far as the tracks are concerned they are fairly uniform. The tracks are too small to be confused with those of the bobcat and are rounder than those of a gray fox or small dogs. Absence of claw marks distinguishes the domestic cat track from that of a fox or dog, and also from mink, which

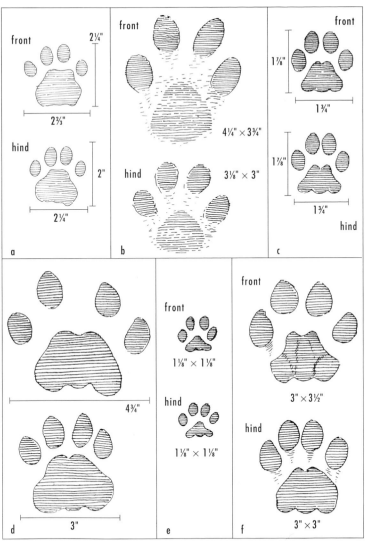

Figure 119. Cat tracks

a. Ocelot.
b. Canada lynx.
c. Bobcat.

d. Jaguar.
e. Domestic cat.
f. Cougar.

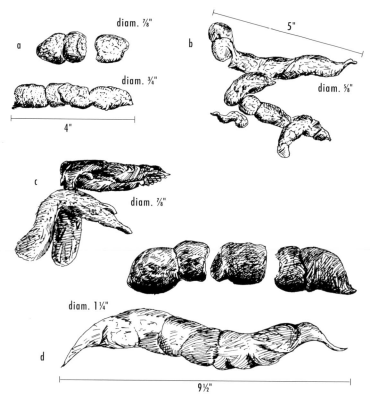

Figure 120. Scats of the cat family

a. Bobcat.   b. Ocelot.   c. Jaguar.

d. Cougar (upper sample, Chisos Mountains, Texas; lower, Olympic Mountains, Washington).

Figure 121 (opposite)

a. Typical tracks.

b. Walking, showing an irregularity in the gait, where the animal paused.

c. Galloping in shallow snow.

d. Walking, and breaking into a bound, in deep snow.

e. Tracks in dust, showing details.

f. Walking, in dust; varying positions of front and hind feet, when they do not register one on top of the other.

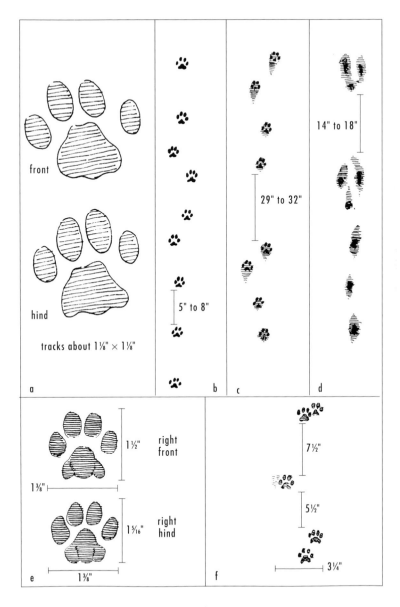

Figure 121. Domestic cat tracks

Front and hind tracks of a feral cat, and an American robin, in southern Wyoming.

shows five toes as opposed to the cat's four. Cats walk neatly and leave a fairly straight trail (see figure 121, b). Front tracks measure 1–1⅝ inches long and ⅞–1¼ inches wide. Hind tracks are 1⅛–1½ inches long and ⅞–1⅝ inches wide.

The feral cat often covers its dung and will seek out suitable substrate in which to defecate. It might be the fact that domestic cats cover their scats so often that leads to the assumption that wild cats cover their droppings just as frequently. But this is not the case.

COUGAR OR MOUNTAIN LION

The mountain lion, or puma, *Puma concolor,* at one time ranged over all of the lower 48 states, within suitable habitats, northward into lower Canada, and in the Northwest, well up into British Columbia, as well as southward through Mexico into South America. Now it is gone from much of this territory.

Figure 122 (opposite)
a. Tracks in mud (Olympic Mountains, Washington).
b. Typical walking trail in snow, showing foot drags.
c. Overstep walk, the common gait.
d. Uncommon straddle trot.
e. Mountain lion wallowing in deep snow.
f. Walking.
g. Another view in snow, leaping gait, showing tail marks (Canada, from photographs by Dr. C. H. D. Clarke).

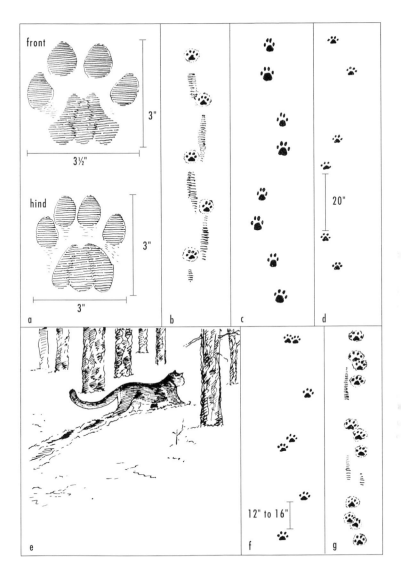

Figure 122. Mountain lion

The perfect hind track of a male Montana cougar.

This American lion, called cougar on the Pacific Coast, is nocturnal and so secretive that the sight of one in the wild is a rarity, and a choice experience. The signs of its presence, then, have added significance. And merely finding the tracks in the snow gives you a thrill second only to glimpsing the animal itself.

The track rarely shows the claws. As with other cats, the front foot is usually wider and the toes tend to spread widely with speed. Front tracks measure 2¾–3⅞ inches long and 2⅞–4⅞ inches wide, while hind tracks are 3–4⅛ inches long and 2⁹⁄₁₆–4⅞ inches wide. The largest measurements will be of splayed tracks in snow. Pumas of the tropics may be expected to produce smaller tracks.

A distinguishing characteristic in some snow areas is reported to be the tail marks in the snow (see figure 1 2 2, g). This is not always present, however, and was not observed in my studies on the Pacific Coast.

As in the case of the bobcat, the droppings of the mountain lion tend to have deep constrictions, or may even be in pellet form in the arid Southwest (see figure 1 2 3). Mountain lions may also cover their droppings with earth, or place them within scrapes created by the hind feet. It is said by "lion" hunters that, when scratching, the mountain lion usually faces in the direction it is traveling. Researchers and hunters also believe that scrapes, and subsequent small mounds of debris, are most often made by male cougars.

It also shares with the wolf and the bears the habit of burying surplus food for future use. One day I was going up Pine Canyon in the Chisos Mountains of Texas. The ground was deeply carpeted with oak leaves. I had stopped at one point to photograph a brightly colored butterfly when I noticed a disturbance in the ground over at one side. I walked over and found a mound of leaves, and under them the remains of a young deer. A little farther on I caught a faint odor of meat and noticed flies buzzing around a certain place, where I found the partial remains of another deer. There was an overabundance of white-tailed deer at that time, and the mountain lion was thus helping to keep them in balance. It was impressive to find such tangible evidence of the presence of this great cat.

Mountain lion dens are generally in caves when caves are available, or in any spot that gives natural shelter.

The voice of the mountain lion has been a subject of controversy. The call has been variously described as screaming, caterwauling, roaring, like the moaning of a woman, yelling, or growling. When we consider the variety of calls of our common house cat, it is easy to understand the diversity in the calls of the mountain lion as they have been described. The few times that I have heard them convince me that they are more or less similar to those of the house cat but magnified many times in volume and depth of tone.

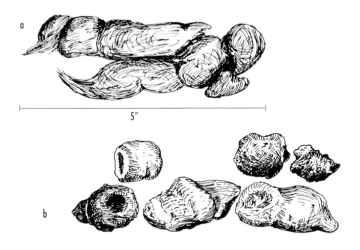

Figure 123. Mountain lion scats
a. From Olympic Mountains, Washington.
b. From Diablo Mountains, Texas.

This beautiful cat, known to science as *Leopardus pardalis*, is at home in the South American tropics but also occurs throughout Mexico and into southern Texas.

The director of the Woodland Park Zoo in Seattle helped me get its tracks in wet sand (figure 124, a). These tended to show the claws, giving a very uncatlike picture. This may possibly be the re-

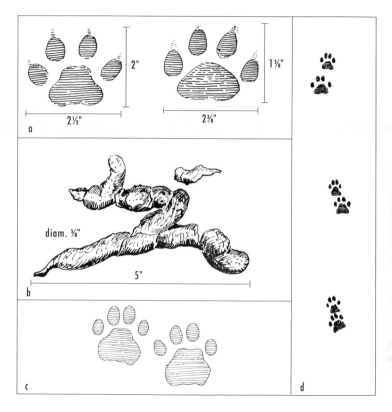

Figure 124. Ocelot tracks and scat
a. Tracks in sand (Woodland Park Zoo, Seattle).
b. Dropping.
c. Front and hind tracks in mud (Mexico).
d. Typical overstep walk of ocelot, in dust (Chamela, Mexico).

Front (below) and hind tracks of an ocelot in dust, in southwestern Mexico.

sult of abnormal claw development in captivity, or of the ocelot's excitement in being driven across the prepared wet sand.

Ocelots in the wild, like other felines, rarely show their claws. In dimensions their tracks are similar to bobcats, and where these animals overlap their tracks could easily be confused in loose substrates or when tracks are splayed. Should you find crisp tracks, there are distinctions to look for in order to distinguish between the two species. The front track of the ocelot can appear more squat— meaning much shorter than wide. Also note the rounder toes of the ocelot, and the blockier palm pad, which does not as clearly illustrate the lobed pads as in tracks of bobcats. Front tracks measure 1⅝–2¼ inches long and 1⅝–2⅛ inches wide. Hind tracks are 1¾–2¼ inches long and 1½–2⅛ inches wide.

The trails of ocelots are much like other felines. The most common walking gait is an overstep walk, where the hind feet step beyond the front track on the same side of the body.

Ocelot scats are much like those of bobcats, although they may contain more lizard and iguana remains.

JAGUARUNDI

This small, slender cat, known as *Herpailurus yaguarondi*, is another product of the South American tropics which has found its way up through Mexico into southern Texas. It is one of those animals, like the screech owl, the red fox, and the black bear, that may be born blond or brunette. That is, there are two color phases —a "red" type, generally rusty brown, and a "gray" type, which is mostly dull gray. In Texas and Mexico, it lives in the arid brushlands. This is one cat that not only climbs trees but readily takes to water as well.

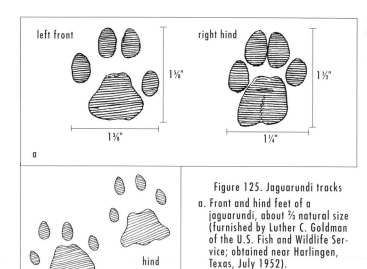

Figure 125. Jaguarundi tracks

a. Front and hind feet of a jaguarundi, about ⅔ natural size (furnished by Luther C. Goldman of the U.S. Fish and Wildlife Service; obtained near Harlingen, Texas, July 1952).

b. Front and hind tracks of a captive female at the Arizona–Sonora Desert Museum in Tucson, Arizona.

Figure 125 shows the imprints of the front and hind feet of a jaguarundi. The second set of tracks were those of a cooperative female held by the Arizona–Sonora Desert Museum in Tucson, Arizona. Front tracks measure 1 1/16–1 3/4 inches long and 1 3/16–1 3/4 inches wide. Hind tracks are 1 1/2–1 13/16 inches long and 1 1/4–1 5/8 inches wide.

Perfect tracks of a wild jaguarundi; photo by Marcelo Aranda.

Lynx canadensis is a close relative of the bobcat, but since it lives primarily in the north country it has adapted itself to cold and deep snow by growing a warm coat of fur, and by having large feet that serve as snowshoes. It inhabits the boreal regions of North America, down into Maine and New Hampshire and along the Cascades and the Rocky Mountains.

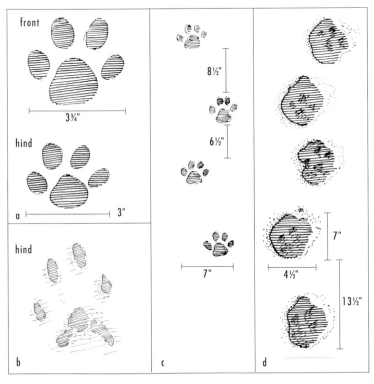

Figure 126. Lynx tracks

a. On hard snow.
b. The hind track of a lynx in shallow mud (Denali National Park).
c. In snow, showing toes; the lynx sank 2 to 3 in. into the snow, and the hind tracks cover the front tracks completely.
d. In deep snow, toes indistinct.

The tracks are larger than those of a bobcat, but could be confused in mud or sand. Yet even in mud, the hind tracks of lynx are larger than those of bobcats. In snow lynx tracks splay and approximate those of the mountain lion, but the lynx does not sink as deeply in the snow as a mountain lion. At times, also, the mountain lion's tail drag shows in the trail. The width of the trail, or straddle, of the lynx represented in figure 126, c, was 7 inches. The straddle of the mountain lion, in the few trails thus measured, has been from 5 to 13 inches. Front tracks measure 2⅜–4¼ inches long and 2⅜–5⅝ inches wide. Hind tracks are 2½–4⅛ inches long and 2⅛–5 inches wide. The largest measurements are in deep snow.

The lynx is, of course, physically much like the bobcat, but has more wide-spreading toes and hairy feet. It is quite customary for lynx kittens to stay with their mother through their first winter, but not so with bobcats. But as cougars may give birth at any time of year, you might also cross the trails of a cougar family in midwinter. I have not been able to distinguish between the droppings of lynx and bobcats.

The lynx is very vocal, indulging in the growls and yelling described for the bobcat, being a much magnified version of the calls of the domestic cat. Various calls have been referred to as "mewing," "yeowing," screaming and moaning, and growling. John Burroughs has referred to one midnight serenade as "a shrill, strident cry, ending in this long-drawn wail."

The trails of a lynx family amidst the crisscrossing trails of snowshoe hare in Glacier National Park, Montana.

The bobcat, or wildcat, *Lynx rufus*, ranges throughout the United States and much of Mexico.

Since it is chiefly nocturnal the bobcat is seldom seen, but its tracks are distinctive. The track is more rounded than that of the coyote or dog and shows no claw marks. When the imprint is imperfect, coyote and dog tracks may not show the claws, either. The ball pad of the bobcat is different from that of the coyote (or any of the dog tribe) in that the anterior edge is two-lobed while the posterior edge is distinctly three-lobed (compare figures 127, a, and 73).

Tracks measure as follows:

front, 1 ⅝–2 ½ in. long, 1 ⅜–2 ⅝ in. wide

hind, 1 ⁹⁄₁₆–2 ½ in. long, 1 ³⁄₁₆–2 ⅝ in. wide

A set of tracks from southern Texas varied. In firm mud they were 2 inches long and 2 inches wide. In softer mud the same feet made tracks 2 ¼ inches long and 2 ⅜–2 ⅝ inches wide.

Apparently the front foot tends to spread the toes more than the hind foot when the animal is speeding. The principal difference between the front and hind tracks is the smaller ball pad of the hind foot. Hind tracks also tend to be slightly longer than wide, though this is not always evident, as it depends on the spread of the mobile toes.

Compare the front and hind (below) tracks of this bobcat in Santa Barbara County, California, with the uppermost track, which is a front track of a coyote.

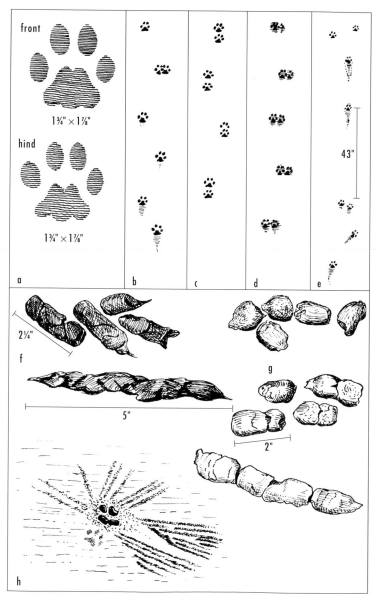

front

1¾" × 1⅞"

hind

1¾" × 1⅞"

a

b

c

d

e

43"

f

2¼"

5"

g

2"

h

Figure 127. Bobcat sign

In walking, the tracks are spaced about 8–14 inches apart. In running, the leaps vary greatly, and may be from less than 4 to more than 8 feet, depending on the urgency of the moment and the depth of snow. As in the case of other winter travelers, in deep loose snow the trail may be a deep trough, with tracks obscured. Look for a favorable location where the individual track is revealed.

The dens may be in a number of places, which vary with the character of the country. They may be in rock crevices or caves, in hollow trees or hollow logs, or in suitable protected places in thickets.

Scats can be confused with those of coyotes and dogs. In the Pacific Coast rain forests they resemble those of the coyote, and this may apply to other humid areas. However, in the arid Southwest the scats tend to become marked off in short segments by constrictions, and sometimes are deposited as pellets (see figure 127, g). This characteristic is useful for identification in these arid regions.

Contrary to the habit of the dog tribe, cats may cover their scats; more often you might find scats dropped in a scrape created by the hind feet. Sometimes the effort is perfunctory, but if the dung is covered, or if there are scratch marks, one may conclude that it is bobcat sign. If the scats are old and scratch marks are obliterated, one must depend on extreme segmentation of the scats for identification, at least in arid regions. Elsewhere, if coyotes are known to be present, mark it doubtful!

Sometimes a bobcat will stretch and scratch a tree, just as the domestic cat will, and as with larger cats, bobcats will cover the remains of kills.

So far as the voice is concerned, again we must say that the bobcat is simply a large pussycat, with the growls, yowlings, and hissing and spitting we are familiar with in the house cat just correspondingly stronger. The first time I heard a bobcat, in Oregon, I was startled by the deep low sound of its growl.

Figure 127 (opposite)

a. Typical tracks.
b. Walking track pattern, stride 9–14 in. (North Dakota).
c and d. Other variations in walking patterns: c, stride 10 in.; d, stride 13–16 in. (Olympic Mountains, Washington).
e. Running pattern (Olympic Mountains).
f. Two scat samples (from Washington).
g. Three samples (Nevada).
h. Bobcat "scratching" (Palm Canyon, California).

I should have liked to see the jaguar in its South American jungle home, with parrots in the trees and perhaps a monkey up among the limbs. In Mexico we would have found it far from the jungle, in the arid brush country and in the mountains. A jaguar track in the dust or on a muddy water margin where we searched would have made an exciting discovery. Instead, my son, with the cooperation of the attendants, obtained the first set of tracks shown here at the Fleishhacker Zoo in San Francisco (figure 128). The second set was of a wild jaguar in southern Mexico.

In general, jaguar tracks are much like those of the mountain lion, and approximately the same size. The principal difference in the tracks of cougars and jaguars is in the shape of the palm pads and the extent to which the lobes on the anterior and posterior edges are visible. Lobes are very defined in cougar tracks and far

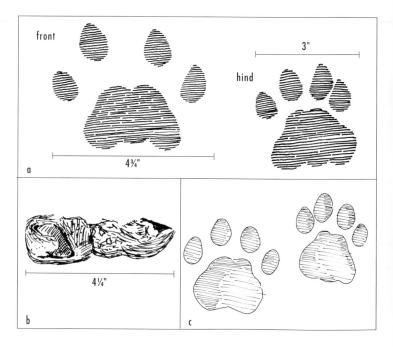

Figure 128. Jaguar tracks and scat

a. Tracks.   b. Dropping.   c. Tracks in mud (Mexico).

The typical trail of a jaguar along a beach in Costa Rica; photo by Marcelo Aranda.

less so in jaguar tracks. Front tracks of jaguars measure 2⅞–3⅞ inches long and 3–4⅞ inches wide. Hind tracks are 2¾–4 inches long and 2½–4½ inches wide.

Jaguars, like cougars, also create scrapes, may cover their scats, and cover their prey for later consumption. According to Jack Childs, jaguars sometimes pull prey down by the nose; thus, damage inflicted to the nose might indicate the kill of a jaguar rather than that of a cougar.

El tigre, as he is known in Mexico, or the *Panthera onca* of science, ranges up through Mexico and has been known to enter Texas, Arizona, and California.

The call has been referred to as a roar, but in a series of short coughlike sounds, and not at all like the roar of the lion. But it is this ability to "roar" that separates "big cats" from "small." Thus the jaguar qualifies for big cat status while our large cougar, which cannot produce such sounds, does not.

SUPERORDER UNGULATA:
HOOFED ANIMALS

THE HOOFED ANIMALS leave tracks that are distinctive as a group, but within the group there are confusing similarities (see figure 129). I would not try to distinguish between the tracks of deer species, for example. Geographical distribution can often be a help in identification. The white-tailed deer inhabits the eastern states, the mule deer the Rocky Mountain region, and the black-tailed deer, a subspecies of mule deer, the Pacific Coast. In the Big Bend National Park in Texas you will find the white-tailed deer on the high forested areas and the mule deer on the more open desert. However, there are intermediate zones in many places where species mingle. It should be noted, too, that although the track of the pronghorn is very often more blocky and squared off behind than those of deer, you will find some tracks that are difficult to distinguish.

Figure 129 (opposite)

a. Mountain goat.
b. Domestic goat.
c. Domestic sheep.
d. Burro.
e. Mule deer.
f. Black-tailed deer (subspecies of mule deer).
g. Mountain sheep.
h. Collared peccary.
i. White-tailed deer.
j. Pronghorn.
k. Pig.
l. Horse.
m. Domestic calf.
n. Elk.
o. Moose.
p. Caribou.
q. Domestic cow.
r. Muskox.
s. American bison.
t. Baird's tapir (right).

t

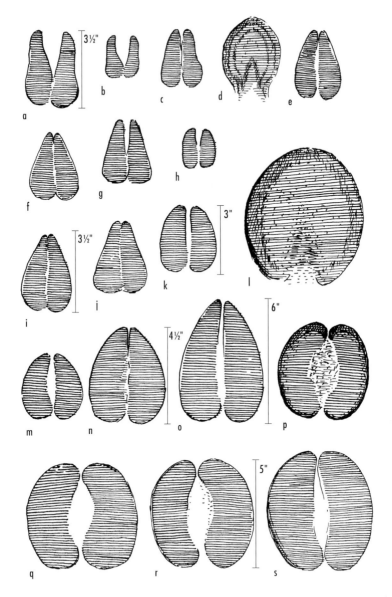

Figure 129. Tracks of hoofed animals, drawn to scale, roughly ¼ natural size

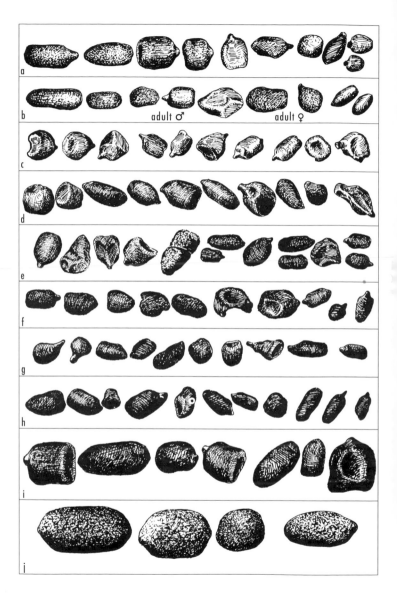

Figure 130. Droppings of various hoofed animals:
pellet type, drawn to scale, about ⅔ natural size

The following keys to droppings of hoofed animals (figures 130 and 131) illustrate the great similarities between the droppings of different species. A series of variations is shown for the pellet types of each animal. There are other variations, however, and it would be difficult to be sure of mule deer pellets in mountain goat, pronghorn, or mountain sheep territory. It is discouraging not to be able to present a more clear-cut key, but that is the way these animals are! So much depends on the kind of food they happen to be eating at a given time, whether dry or succulent or ranging in between.

Figures 130 and 131 give both the pellet type of droppings that characterize the winter feeding or a dry diet, and the soft type of droppings that result from green or succulent food in summer. Here, too, there are many variations in appearance and size, only a few of which can be given.

For more detailed treatment see the sections devoted to each animal that follow.

HORSE AND BURRO
(ORDER PERISSODACTYLA, FAMILY EQUIDAE)

Horse tracks can hardly be confused with anything else. The horse wrangler in the western mountains frequently has occasion to hunt up his saddle and pack animals when they stray from camp, and since he must often rely on fragmentary tracks, his knowledge of the horse track is necessary. Typically, the horse track reveals the single round or oval hoof, with the V mark of the frog in the middle. A shod horse, of course, will make a track outlined rather strongly with the iron rim; in firm mud it shows the toe and heel calks. Mule tracks are smaller and narrower than those of a horse, and burro tracks are smaller still. The tracks shown in figure 132, a and c, are those of unshod saddle horses, the type one will find in the mountains. Obviously, the large work

Figure 130 (opposite)

a. White-tailed deer.
b. Mule deer.
c. Caribou.
d. Pronghorn.
e. Mountain sheep (first 3 samples from bighorns in Wyoming; next 5, desert bighorns; last 5, Dall's sheep in Alaska).
f. Domestic sheep.
g. Mountain goat.
h. Domestic goat.
i. Elk.
j. Moose.

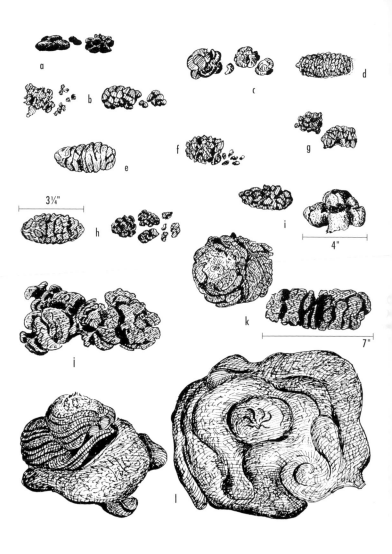

Figure 131. Droppings of hoofed animals, of the soft summer type, drawn to scale. It is important to note that there is great variation in form for each species. These are only random specimens. In some cases 2 samples are illustrated. A few measurements are included for estimate of size. Note that domestic cattle droppings are similar to those of bison.

horses on farms, if any remain in this mechanized age, make much larger tracks.

Horse droppings are distinctive and familiar. There is an interesting observation that probably harks back to the days when wild horses first roamed the western plains many centuries ago. We once more have some wild horses in Wyoming and a few other western states, and wild burros in the Southwest. Among these wild horses it has been observed that when a stallion has established a local home range with a band of mares, it will leave droppings in the same spot repeatedly, until a considerable pile has accumulated. A male African rhinoceros will do the same. In view of this it is interesting to speculate that, after all, horse and rhinoceros are related, both being odd-toed ungulates!

In some places in the West it is still possible to hide at a water hole and see at the break of day the pronghorn and wild horses come in through the sage for a drink of water. It is a stirring sight when a band of horses strings along down a desert ridge, manes tossing, eyes looking here and there alertly—once again the wild animals that they were long before known human history.

DOMESTIC PIG AND WILD BOAR
(ORDER ARTIODACTYLA, FAMILY SUIDAE)

Pig tracks may be found near highways in the Middle West, but may also be found in some areas where they have gone wild, or where they forage in deer country. The wild boar of Europe, *Sus scrofa*, has been introduced into the United States, notably in the regions in and around the Great Smoky National Park, as well as in Florida, Texas, and coastal California.

The tracks of domestic pigs, along with those of the wild boar, are more rounded than those of deer. Also, the pig and boar dewclaws are more pointed and extend farther out to the side (see figure 133, b). Measurements for wild boar tracks are as follows and do NOT include the dewclaws. Front tracks measure 2⅛–2⅝ inches long and 2¼–3 inches wide. Hind tracks are 1⅞–2½ inches long and 2–2¾ inches wide.

Figure 131 (opposite)

a. Collared peccary.	g. Domestic goat.
b. Mule deer; 2 samples.	h. Caribou; 2 samples.
c. White-tailed deer.	i. Pronghorn; 2 samples.
d. Mountain sheep.	j. Moose.
e. Domestic sheep.	k. Elk; 2 samples.
f. Mountain goat.	l. American bison; 2 samples.

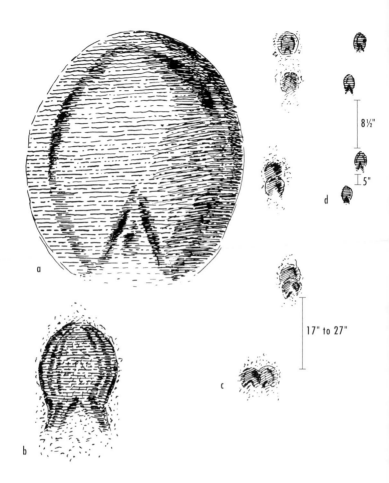

Figure 132. Horse and burro tracks

a. Wyoming saddle horse track in snow. Front track was 6 in. long, hind track, 5½ in.

b. Young burro track in dust, about 2½ in. long (Nevada).

c. Track pattern of a saddle horse in light snow. Note the variations in the position of front and hind tracks. In this case the front track was forward or even with the hind track.

d. Track pattern of young burro shown in b.

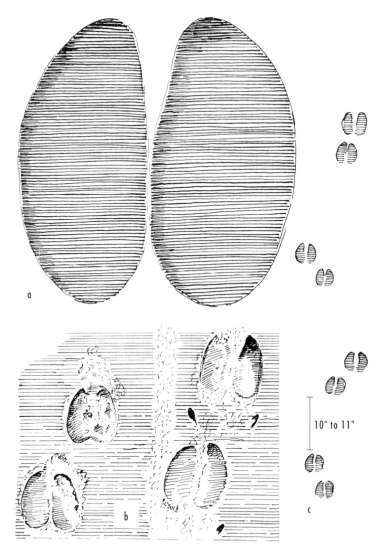

Figure 133. Tracks of domestic pig
a. Adult track in mud, about ⅔ natural size (Iowa).
b. Tracks in snow (Wyoming).
c. Trail of adult sow, in mud (Iowa).

10" to 11"

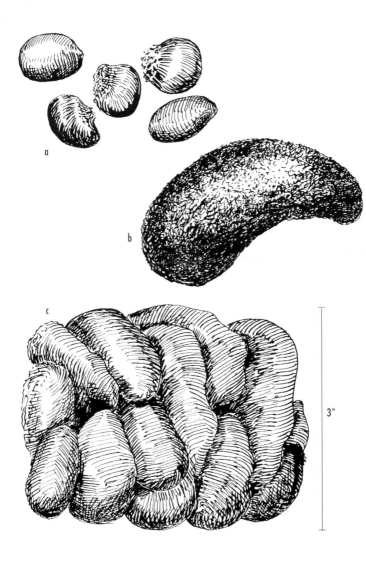

Figure 134. Droppings of domestic pig from a Pennsylvania farm, about ⅔ natural size

a. Pellet form.   b. Unsegmented type.
c. Common type in which pellets are retained in a mass.

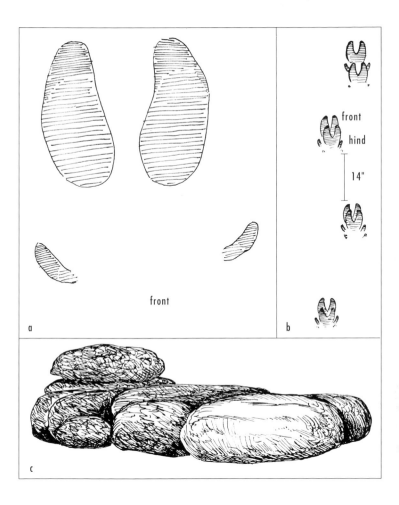

Figure 135. Wild boar tracks and scats

a. Track of front foot (Texas). Hind feet are slightly smaller than front feet. Weight of animal causes the hooves to spread. Dewclaws of front feet usually show in track but seldom those of hind feet.

b. Walking trail, with hind feet being placed in or partly over impression of front feet (from Blue Mountain Forest and Game Preserve, Newport, New Hampshire; drawn by Carroll B. Colby). In traveling, boars usually trot, where strides then measure 26–32 in. between tracks.

c. Droppings of wild boar vary greatly with diet (drawn by Carroll B. Colby).

The wild boar is of course more agile than the domestic pig, with narrower hooves and longer legs and strides. The wild boar, domestic pig, and peccary all root up the earth and more or less wallow, leaving signs of this activity in the earth. The wild boar also rubs on the trunks of trees and telephone poles, and gouges them with its tusks. In all such places, as in the case of bear trees, one should look for telltale hairs and tracks (figure 135).

Mr. LeRoy C. Stegeman, who has studied the wild boar in Tennessee, tells us that wild boars have been known to rub trees to a height of 37 inches. He comments:

> The trail of a wild boar is narrower than that of the domestic, the tracks being made almost in a single line. The tracks of the domestic hog are offset, forming two lines or a single zigzag line [see figure 133, c]. The wild boar will run or jump up banks too steep and high for the domestic hog to climb. The wild boar is a good jumper and will leap over obstacles such as down logs, that the domestic would go around or under. Wild boars frequently cross streams by walking across down logs, while domestics would go through the stream. Wild boars are much taller than domestic swine. Therefore, the height to which trees are muddied may be good evidence of identity.

Domestic pig droppings are variable in form, depending on the kind of food eaten. They may be soft and almost formless, but on more solid food they tend to form in pellets, either coalesced into one mass, as in figure 134, c, or separate as in a. In b of this figure is a type of a more uniform consistency without segmentation, which might be confused with bear scats. When feeding on grass the domestic pig produces pellets that are dry and have the appearance of miniature horse dung. Thus we find in the pig a type quite comparable to those of the elk and moose, with their seasonal variations.

COLLARED PECCARY OR JAVELINA (ORDER ARTIODACTYLA, FAMILY TAYASSUIDAE)

The collared peccary, known in the Southwest as the javelina and in science as *Pecari tajacu,* is related to the pig family. Its home is in Central America and Mexico, but it has also established itself in southwestern Texas and southern Arizona, as well as in southeastern New Mexico.

The peccary inhabits the mesquite and cactus country, sometimes in sparsely wooded foothills, also high up in some of the southern mountains. It is known to use dens in the ground or hollow logs, sometimes caves in the rocky cliffs. In one trip to Arizona I found piles of droppings in such caves. Most were like those in figure 136 but there were also many flat ones, like a rough pancake, deposited when semiliquid. The food of the peccary consists of a great variety of things, including roots, fruit, cactus, nuts, insects, eggs, and any flesh that can be procured, and the shapes of their scats are just as varied.

The hoof marks are much smaller than those of pigs but are of the same general pattern (see figure 136). The peccary has only one dewclaw on the hind feet, but it does not show in the tracks unless the animal is bounding. An active animal, the peccary will readily leap a distance of 6 feet, and has been recorded as leaping as much as

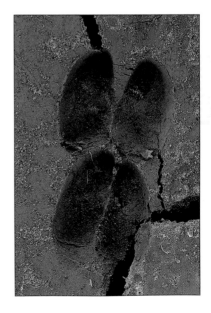

Perfect front and hind tracks of a collared peccary in Big Bend National Park, Texas.

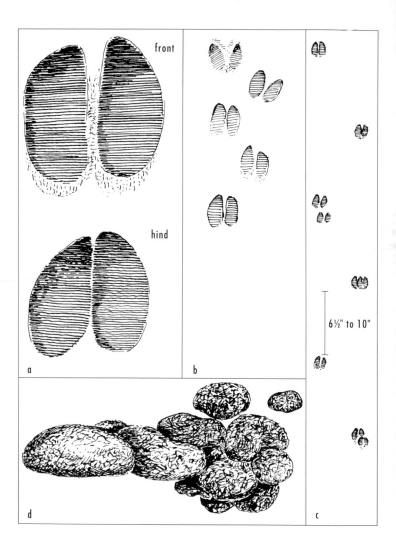

Figure 136. Peccary tracks and droppings

a. Typical tracks in clay, natural size. Front hooves about 1½ in. long; hind, 1¼ in. (Fleishhacker Zoo, San Francisco).

b. Tracks in mud, showing variations (Texas and Arizona).

c. Walking pattern, in mud. Width, or straddle, about 4–5 in. (Arizona).

d. Dropping (Arizona).

10 feet 9 inches when startled. Front tracks of the collared peccary measure 1 –1 ⅞ inches long and ¾–2 inches wide. Hind tracks are 1 –1 ⅞ inches long and 1 ¹⁄₁₆–1 ¾ inches wide.

When startled sufficiently or excited in any way, it produces a strong scent definite enough to be detected by the human nose and probably easily detected and interpreted by other peccaries or by enemies.

These animals express themselves with a grunt or squeal. Their presence is most readily detected, however, by their rooting places in the earth and by their tracks, or by the piles of scats along their runs or in caves. You will also likely see them moving about in small to medium bands during the day if you travel along or wait next to their well-beaten paths.

ELK, DEER, MOOSE, CARIBOU, AND REINDEER
(ORDER ARTIODACTYLA, FAMILY CERVIDAE)

ELK OR WAPITI

The elk, *Cervus canadensis,* is now being redistributed into many parts of the continent where it was once numerous. Small populations now exist in Michigan, Texas, North Carolina, Pennsylvania, and Oklahoma, as well as large populations throughout the northern and southern Rockies, and the coast of the Pacific Northwest.

Elk tracks are definitely larger and rounder in outline than those of deer, and there is

The hind track of an elk overlaps the front track in Grand Teton National Park, Wyoming.

no difficulty in distinguishing them. They are rounder and some-what smaller than those of moose, though occasionally even these may cause a little confusion. In localities where elk share the same range with Hereford cattle there can be some difficulty. The adult cattle tracks are large and blocky, and quite distinct. But *young* cattle leave a track that is very similar to those of adult elk. Look closely at the shape of the hoof wall between the cleaves—the elk's are straight, the cow's curved—as well as at the length of the stride and other associated signs to differentiate the two species. Tracks are shown in figure 137, which includes a track of a domestic calf for comparison.

Front tracks measure 3–4⅞ inches long and 2⅜–4⅝ inches wide. Hind tracks are 2½–4½ inches long and 2⅜–4 inches wide.

In New Zealand it was noted that wapiti living on moist lowland earth developed exceptionally long and pointed hooves, while those on the rocky uplands had worn the hooves relatively blunt. It is well to keep this principle in mind in studying hoofprints.

The elk scats are fairly distinctive, but examination of figure 139 proves that variability is great. In joint elk-moose country some puzzling samples will be found, for the elk, too, will produce elongated "sawdust" pellets somewhat similar to those of moose, when on a dry, exclusively browse diet. However, the quantity of the scat material is much less in the elk.

In summer the droppings lose the pellet form and are left as flat, elongated or circular chips similar to those of domestic cattle. Elk chips are smaller, usually about 5 or 6 inches in diameter, though they may be longer. The much larger cow chips are at least twice as big. Of course, in either case there may be very small ones.

The sign left by European red deer, *C. elaphus*, which have been imported to America, is very similar to that of elk, though the average tracks are smaller.

Figure 137 (opposite)

a. Adult cow elk track, on sandy soil (Wyoming).

b. Adult elk track (Olympic Mountains, Washington).

c. Adult elk track (Wyoming).

d. Domestic Hereford calf track, for comparison (Wyoming).

e. Elk track, showing dewclaws when running (Wyoming).

f. Elk track in New Zealand, showing the long pointed type when animal has been living on soft ground.

g. Elk calf track (Wyoming, June 13, 1932).

h. Aspen trunk, with elk tooth mark, and elk barking aspen in a "high-lined" grove.

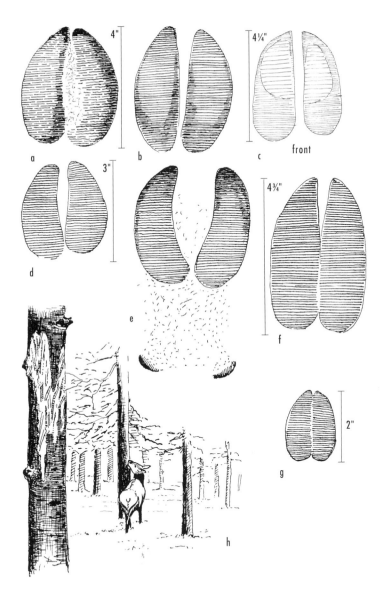

Figure 137. Elk tracks and browsing

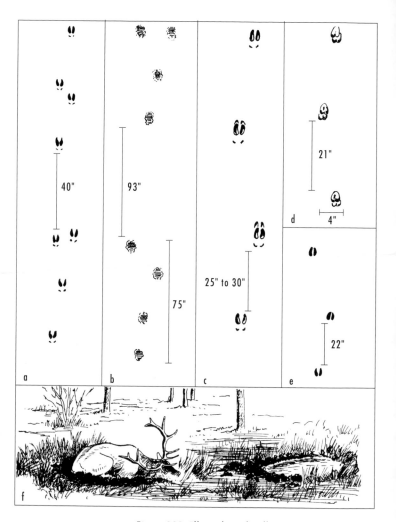

Figure 138. Elk tracks and wallow

a. Cow elk, bound into gallop, in mud (Wyoming).
b. Elk gallop into bound in snow (Wyoming).
c. Cow elk walking in sand (Olympic Mountains, Washington).
d. Yearling elk, walking; hind foot slightly forward of front track (April 9, 1928).
e. Young calf elk, walking (Wyoming, June 1, 1929).
f. Bull elk wallowing, during the rut; an old wallow, into which water has seeped, shown at right.

There are other signs of elk. In winter you will find pits dug in the snow where elk have been pawing through for food, or you may find the beds in the snow, or in summer the flattened places in the grass, where they have rested.

I remember a particularly pleasant day on summer elk range in the Rockies. Over a rise I came upon a depression, a green meadow, in the middle of which was a pond. A band of elk was just then coming in to water. As they came near, some of the eager ones rushed on ahead, jumped into the water, and romped along in the shallows, splashing the water, shaking their heads, and hopping with the same joy that a group of children go splashing into the water at a beach. This is one of the rewards of watching the animals in their native home.

If you are in high-mountain elk country and come upon a shallow pond, look for tracks at the muddy borders. If the water is roiled up and milky you may guess that elk have been there to bathe, drink, and frolic not so long ago.

In the autumn during the rutting season, if you don't actually find a bull elk prodding the wet ground with his antlers, pawing the dirt with his hooves, and wallowing in the mud, you should at least find the muddy pits where the bull elk has cooled his mating fervor. Sometimes you will find a permanent little pool of water, with vegetation around it, which was an elk wallow several years before, into which water has seeped (see figure 138, f).

The elk will also thrash about in bushes and small trees with his antlers to spread his scent. Saplings may have limbs broken off and the bark may be torn and bruised. Often saplings are girdled and die, and you will find many such dead trees in a given area, victims of the ardor of the wapiti rut. On such trees the bark is obviously bruised, torn, and shredded by the rubbing, in contrast to those that have been gnawed.

There is still another tree sign. Elk are fond of the bark of certain trees, particularly the aspen, and you will often find the bark scored by tooth marks where the animal has scraped with its lower incisor teeth. These grazing animals have no upper incisors. This is done also by moose and deer, and one must guess which it is by knowledge of which animal is known to spend the winter in a given locality. Aspen trunks that have been gnawed year after year eventually develop a rough, blackened trunk as far up as the elk reached. Such a black-trunked aspen grove denotes a much-used elk winter range.

The mineral lick is also worth noticing. Here the ground is gouged out by the mud-eating animals—deer, elk, moose, or mountain sheep. The tracks in the mud will tell you which. In the forests of the Olympic Peninsula you will find old hemlock logs gnawed by elk, the lower sides apparently being preferable. I have

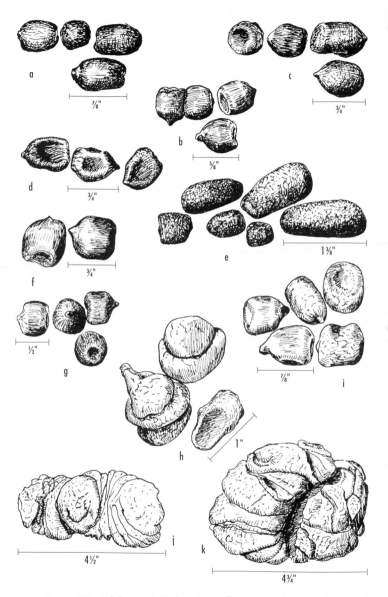

Figure 139. Red deer and elk droppings; all except j and k about ⅔ natural size

seen pits gouged out under such a tasty log where the animals had knelt to get at the underside. Sometimes, too, you will find old rotten limbs or small tree trunks, mushy in consistency, eaten into by the elk.

Finally, there is one delightful indication: the bugle call in the autumn months. If you are fortunate enough to be in elk country at that time of year, a moonlight night can be specially enriched by the calling of the bull elk. It rises with a glide to a high-pitched silvery note, then glides down again, to end in some guttural grunts. This is wapiti music.

There are other sounds. The mother and calf will call to each other with a squealing sound, more pleasing to hear than this description implies. When a band has been put to flight by some disturbance, then slows up to reassemble, there may be a babel of voices as the mothers and calves seek each other.

There is also a sharp, loud bark. This explosive sound signifies either alarm or curiosity, perhaps a combination of the two. I have heard an elk bark repeatedly for a long time, evidently puzzled by a lighted tent at night.

All such vocal expressions enliven your experience with the elk herds in their native surroundings.

WHITE-TAILED DEER

When we speak of deer, to most Americans it probably means the white-tailed or "flagtail," as it is often called, *Odocoileus virginianus*. This is the common deer of eastern and central North America. Its range extends northward into southern Canada,

Figure 139 (opposite)

a and b. Adult red deer stag droppings (southern New Zealand, March 28 and April 5, respectively, 1949).

c. Adult female red deer (southern New Zealand, March 24, 1949).

d. Droppings of elk bull in its 4th year (Wyoming, October 25, 1950).

e. Roosevelt elk droppings, a woody browse diet (Olympic Mountains, Washington, March 1935).

f and g. These two samples are from American elk introduced in New Zealand, both large bulls weighing over 550 pounds (in the same locality, early March 1949).

h. Soft pellet type (Sun River, Montana, July 21, 1934).

i. Common pellet type (Sun River, Montana, July 21, 1934).

j and k. Soft coalesced type, from most succulent forage.

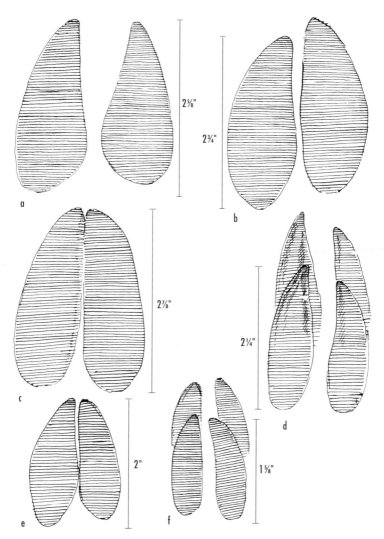

Figure 140. White-tailed deer tracks in mud

a, b, c. Various tracks from Wichita Mountains, Oklahoma.

d. Tracks from northern Minnesota (1924).

e. Track of a small deer in Chisos Mountains, Texas (March 1950).

f. Fawn tracks from Michigan (July 7, 1934).

southward through Mexico and Central America, and westward into the Rocky Mountain region, including parts of Idaho, Montana, Wyoming, Colorado, New Mexico, and southern Arizona. And there are a few small colonies in Washington, Oregon, and California.

Generally speaking, this is the deer of the forests and brushlands, in contradistinction to the mule deer of the West, *O. hemionus*. As might be expected over such a wide and varied range, there is much diversity in size, from the large deer of Minnesota and Wisconsin, which weigh several hundred pounds, to the smaller, more trimly built deer of Texas and Mexico. The "Key deer" of certain Florida Keys is the smallest of all. The few survivors of this diminutive subspecies are struggling for existence in the face of an expanding real-estate business. This striking diversity in size among the white-tailed deer is reflected in their tracks and other sign. In figure 140 are shown tracks from Minnesota, Oklahoma, and Texas. Those of the Key deer are even smaller than those from Texas.

The front tracks of white-tailed deer measure 1⅜–4 inches long and ⅞–2⅞ inches wide. Hind tracks are 1¼–3½ inches long and ¾–2⅜ inches wide.

When we come to figure 141, showing track patterns of this deer, we should note a conspicuous difference between the actions of the white-tailed and the mule deer. In galloping, the white-tailed uses the "rocking-horse" gait so common among large animals, in which the hind feet swing far ahead of the front-foot tracks. This produces the familiar hind-tracks-beyond-front-tracks pattern so common, from the white-footed mouse and rabbit to the moose and horse, in running gaits. The mule deer, on the other hand, often proceeds in speed

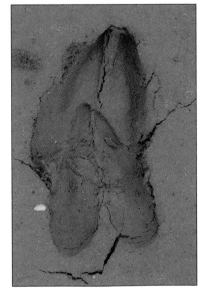

Sharp front and overlapping hind tracks of a white-tailed deer in coastal Texas.

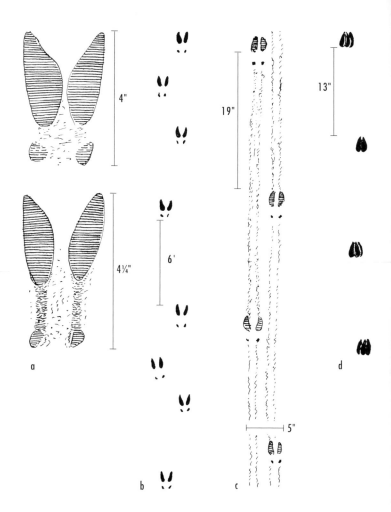

Figure 141. Track patterns of white-tailed deer

a. Leaping tracks, in mud, with dewclaws showing. In front track (upper) dewclaws are close to hoofs. In hind track (lower) dewclaws are farther from
hoofs (Oklahoma, 1935).

b. Galloping track pattern, in snow; hind tracks in front (Michigan).

c. Walking pattern in snow, showing drag marks of toes (Michigan).

d. Walking pattern of young deer, on dirt road, showing the traditional heart
shape of footprint (Minnesota).

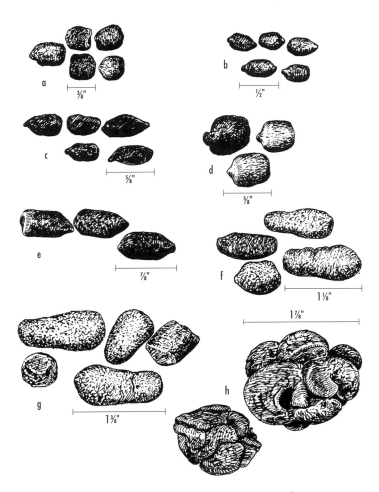

Figure 142. White-tailed deer droppings, about ⅔ natural size

a. A smaller type (Minnesota).

b. Pellets (Chisos Mountains, Texas).

c. Softer type of pellets (Minnesota).

d. Winter droppings (Wisconsin).

e and f. Large types (Minnesota).

g. Unusually large winter droppings (from a deer yard near Grand Marais, Minnesota).

h. Soft summer droppings (Minnesota).

with a bounding "rubber ball" action, all four feet coming down to-gether, hind feet behind. The term for this gait is a *pronk*.

You will notice among the tracks of this and the other deer that you do not always find the neat, well-formed, heart-shaped print so often figured. Very frequently the hooves are separated. More-over, I have been unable to distinguish the footprints of the white-tailed from those of mule deer. Fortunately, over most of the range only whitetails are present. Don't worry about confus-ing any animal with mule deer in the Adirondacks, the deer coun-try of Pennsylvania, Minnesota, or indeed over most of the coun-try east of the Rockies, for there are very few mule deer in these parts. In areas where both species occur, the white-tailed deer generally inhabit moister terrain than the mule deer, though this is just a tendency and not a hard-and-fast rule.

Like the elk, the whitetail buck will occasionally wallow in mud during the rutting season. Such messed-up muddy spots, if made by deer, should contain some identifying hair. However, this is quite rare. But bucks do often create scrapes during the rut along their travel routes, in which they urinate. Look also for frayed saplings where deer have rubbed their head and antlers in prepa-ration for mating rituals.

Should you see a white flashing signal disappear before you at dusk, or in heavy brush, it may have been a fleeing whitetail. For when a whitetail runs, true to its name it holds aloft its large white tail, which becomes a prominent sign that can be seen at considerable distance.

The deer too have a voice. Most commonly heard is a sharp snort or whistle. This is a somewhat prolonged *whiew-ew-ew,* which carries quite far and is given when the deer is alarmed or curious. If you are close enough, deer can also be heard to stamp when curious or nervous about something they cannot identify. There is in addition a sort of bleat, a sound produced by the vocal cords that is difficult to describe. The low calls between fawn and mother are hardly to be detected at any distance.

MULE AND BLACK-TAILED DEER

The mule deer of the West, *Odocoileus hemionus,* is larger than the whitetail, has a different antler structure, and has a round "mule tail" tipped with black, in contrast to the large white flagtail of the eastern deer. The mule deer occupies western America from the northwestern tip of Minnesota and southwestern Manitoba, the Dakotas, westernmost Nebraska, Colorado, New Mexico, western Texas, and northern Mexico west through the Rocky Mountain re-gion to eastern British Columbia, eastern Washington, Oregon, and California. It also extends far north in the Canadian provinces.

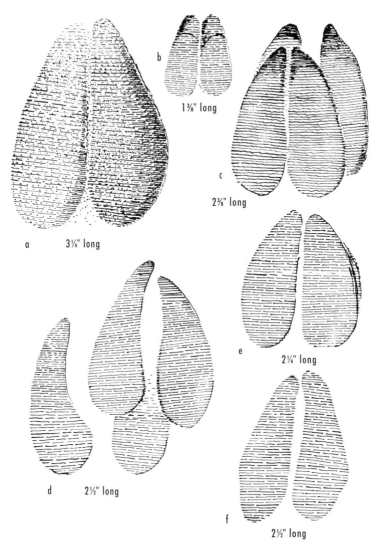

Figure 143. Mule deer tracks, about ⅔ natural size
a. Adult male, in dust (Grand Teton National Park, Wyoming).
b. Fawn (Yellowstone National Park, July 16, 1929).
c. Adult female, in mud (Grand Teton National Park).
d, e, f. Tracks in dust (northern Nevada).

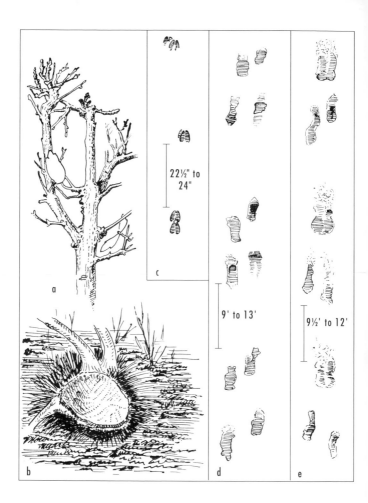

22½" to 24"

c

9' to 13'

9½' to 12'

a

b

d

e

Figure 144. Mule deer sign

a. Shrub severely browsed by mule deer, producing a stubby form.

b. Base of the spiny lechuguilla eaten out by mule deer (Chisos Mountains, Texas).

c. Walking track pattern (Wyoming).

d and e. High-bounding, or pronk, track patterns in deep snow of adult female mule deer. In d the doe was pursued by a coyote. In these patterns the hind feet land behind the front feet.

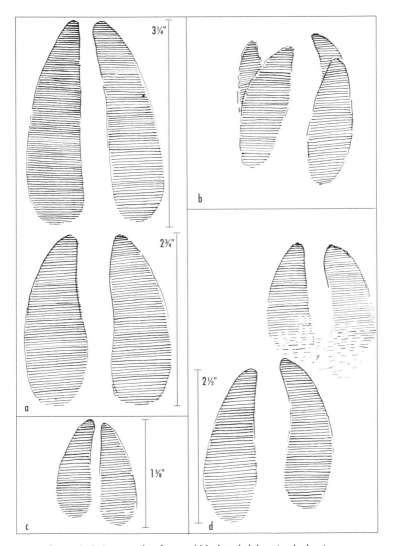

Figure 145. Some tracks of coastal black-tailed deer (mule deer)
a. Adult doe, in mud (Olympic Mountains, Washington, March 1934).
b. Yearling, accompanying this doe.
c. Young Sitka deer (Long Island, September 3, 1936).
d. Adult Sitka deer (Long Island, Kodiak, Alaska, September 3, 1936).

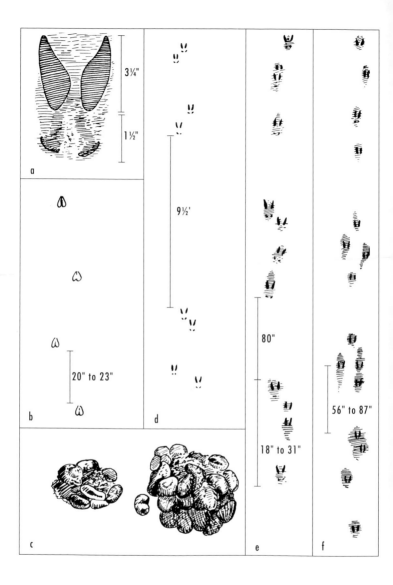

Figure 146. Black-tailed deer subspecies tracks and droppings

This deer often ranges into more open and drier country than the habitat chosen by the whitetail. It is true that mule deer inhabit the western forests, but they are more of a mountain animal than the whitetail, and in many places are found out on the desert plains, in both cactus and sagebrush country.

The black-tailed deer of the Pacific Coast was once classed as a distinct species but is now considered a subspecies of the mule deer, with which it shares many characteristics, including the same antler type and even the same bounding gait. But it has a flat tail much like that of the white-tailed deer, yet smaller and black on the upper surface. It is found from the Sierras and Cascades west to the coast in California, Oregon, and Washington, and northward along the coast of British Columbia into the southeastern coastal part of Alaska.

As described in the whitetail section, the mule deer has a distinctive bounding gait called a pronk, in which all four feet virtually come down together, hind feet behind the front feet. This track pattern is shown in figure 144, d and e. Figure 143 illustrates a variety of footprints.

The shape of the track of the mule deer will vary somewhat with the type of ground on which the animal lives. On soft soil, as in some woodlands, the toes are likely to be relatively more pointed. On hard rocky ground, found in some areas occupied by these deer, the hooves are worn enough to produce blunt tips.

Deer sign found in the coastal rain forests will be of the black-tailed subspecies, but in parts of the Cascades and Sierras the ranges of mule deer and coastal black-taileds overlap. I must confess that I have found no way to distinguish the footprints or droppings of our deer species. Black-tailed sign is shown in figures 145 and 146. Tracks will reveal gaits similar to that of mule deer.

During the rut, mule deer produce similar signs to white-tailed deer, as described above.

Figure 146 (opposite)

a. Typical track at high speed, with the dewclaws registering well in the snow (Olympic Mountains, Washington).

b. Walking pattern, with conventional track shape, toes together (Olympic Mountains).

c. Sitka deer droppings, soft type, about ⅔ natural size (Hinchinbrook Island, Alaska).

d. Bounding track of yearling, toes spread (Olympic Mountains, March 24, 1934).

e. Galloping gait track of female, in shallow snow (Olympic Mountains).

f. Same, in deep snow (Olympic Mountains).

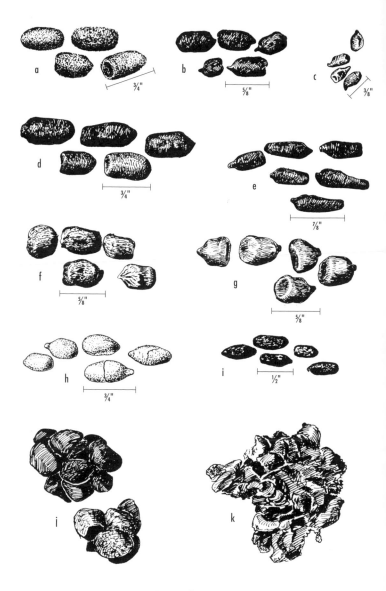

Figure 147. Mule deer droppings, about ⅔ natural size

The walking trail of a mule deer in sand dunes of western Texas.

As with all deer, the droppings vary in size and shape, as shown in figure 147. The quantity in a given sample varies from less than ⅛ to ½ pint, and the number of pellets in a series that was studied varied from 68 to 128. Incidentally, I have been unable to distinguish with certainty between some deer pellets and those of the mountain goat.

The mule deer emits a snort, or blowing, sometimes prolonged slightly to produce a whistling effect. This may indicate alarm or surprise, or sometimes curiosity. Like other deer species, as well as the pronghorn, the mule deer will stamp a front foot. Is this defiance, fear, or excitement? All that we can say is that it denotes some kind of intense feeling.

The mother and fawn, of course, make the low bleating sounds common among such animals. A neighbor of mine who lives on a ranch beside a high hillside extending up to the forest told about hearing the mule deer "talking" among themselves as they fed

Figure 147 (opposite)
a. Winter droppings of adult male (Dinosaur National Monument, Utah, 1950).
b. Autumn droppings of adult male (Wyoming, November 8, 1938).
c. Fawn droppings (Wyoming).
d. Adult female droppings (Wyoming, January 22, 1939).
e. From Wyoming (April 1937).
f. From British Columbia (June 12, 1935).
g. From northern Nevada (August 24, 1939).
h. Droppings of pure clay, resulting from use of mineral lick (Salmon National Forest, Idaho).
i. Scats probably from young animal (Wyoming).
j and k. Summer scats, of the softer type.

there on winter nights. They made various sounds, grunts, and other expressions, difficult to describe. These deer offer excellent opportunity for study of vocal performance of animals and their meanings.

MOOSE

This is our largest deer, *Alces alces,* which inhabits the boreal forests of North America and the northern part of the Rocky Mountains. Moose are found from Connecticut north in the Northeast as well as in northern Wisconsin, Michigan, and Minnesota. There are also rare sightings in northern New York. In the Rocky Mountains they are present chiefly in western Montana, northern and central Idaho, and western Wyoming. Moose occur all the way across the wooded portions of Canada, throughout most of Alaska, and even out on the Alaska Peninsula. Willow growth appears to be closely associated with moose distribution over most of its range, though moose feed on a variety of wetland plants and woody vegetation.

The moose track is nearest to the elk track in size and shape, though it is larger and more pointed (figures 148 and 149). Occasionally a moose with blunter hooves will produce an unusually rounded track. One day when I was out in the mountains, a veteran skier and mountain climber asked me, "How do you distinguish moose and elk tracks?"

I explained the difference, stressing the more elongated, sharper-toed form of the moose track. A closer study reveals a significant difference in the pads on the feet. Elk tracks have soft pads only at the posterior edge of tracks, while the pads on moose feet run the en-

Deep-set tracks of a cow moose and a young calf in central Alaska.

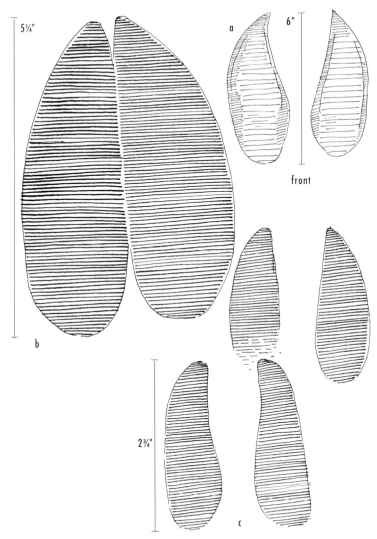

Figure 148. Moose tracks, about ⅔ natural size

a. Adult moose track in mud (Alaska).
b. Adult track in mud (Wyoming).
c. Calf front and hind track in mud (Wyoming, June 30, 1951).

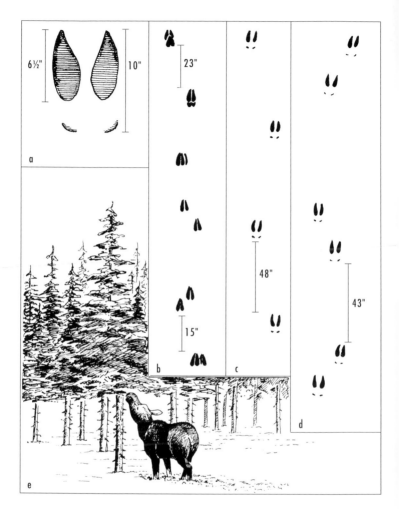

Figure 149. Moose tracks and browsing

a. Track in running pattern, with dewclaws showing (Alaska).

b. Irregular walking pattern (British Columbia, 1935).

c. Adult cow moose trotting, on old hard snow (Wyoming, 1950). A walking pattern of another cow moose in the same locality was similar, but with steps from 24 to 30 in.

d. Another fast trotting pattern (Alaska, 1922).

e. A "high-lined" grove of fir trees, on which foliage has been eaten as high as the moose can reach (Wyoming).

tire length of the hooves. Compare the illustration in figure 137, c, with figure 148, a. Front tracks of adult moose measure 4⅜–7 inches long and 3¾–6 inches wide. Hind tracks are 4⅛–6½ inches long and 3½–4⅝ inches wide.

Moose droppings in winter are distinctive. Because of the dry browse diet in that season they appear like compressed sawdust, and may be round or elongated, usually smooth. The variations are shown in figure 150. The quantity of pellets in a sample is greater than that of an elk, as would be expected from such a large animal. In 18 samples of winter droppings measured, the average quantity was 1 quart, with a minimum of ½ quart and a maximum of 2 quarts. The number of pellets varied from 78 to 192, with an average of 128. One fact to be noted is that a sample found at a moose bedding place, where the animal has been lying a long time, is likely to be large, while the samples along the route of travel, where a moose has been feeding, will be more frequent and less in quantity.

In midsummer the droppings are soft and a sample may be almost formless, resembling certain ones of the domestic cow. In spring and fall, with the shift to and from a succulent diet, the samples are intermediate, retaining the pellet form but distorted and clinging in a mass. The total mass will be greater than that of an elk.

You will find moose beds in snow or in tall grass where the vegetation has been mashed down. Since there is such a great diversity of size, from the calf and yearling to the large bull, it would be difficult to distinguish a moose bed from an elk bed by size alone in an area occupied by both animals. Nearby tracks and droppings are a clue.

In winter moose will trim the limbs of fir trees, aspens, and other favorites as high as it can reach. Traditionally, the moose is recorded as "riding down" a sapling by bending it over with its body and straddling it so as to reach the top twigs. I must admit that I have never seen moose do this, but I have seen them reach high up, seize the stem in their teeth, and bend it over until it breaks. Then they will proceed to feed on the drooping tips. I have seen them do this with young cottonwoods and tall willow brush. A much used willow winter range will be characterized by numerous broken tops, dead and slanting across or downward in diverse directions, as shown in figure 151. Moose will also feed on the barks of aspen, red maple, and balsam fir trunks, as shown in figure 152, a.

In summer moose feed extensively in ponds, where they seek the submerged aquatic vegetation. They have favorite ponds that they visit repeatedly, especially in early morning or in the evening. They do not graze extensively. Their necks are too short for reach-

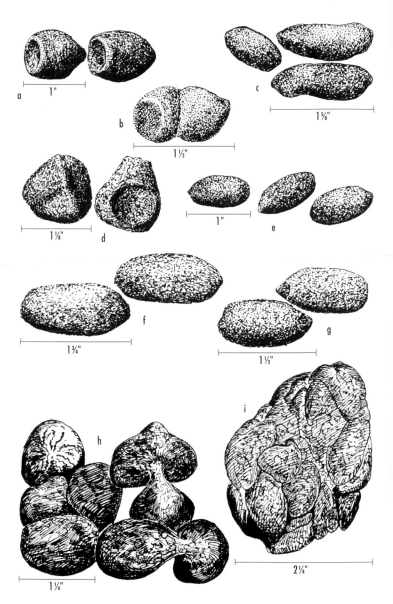

Figure 150. Moose droppings, about ⅔ natural size

Figure 151. Moose browsing on willows in winter, showing willow tops broken over to bring the high twigs within reach (drawing by Grant Hagen)

ing the grass easily, but I have seen them get down on their front knees to do so, and they do have a number of other herbaceous plants in their diet. One day we watched a cow moose busily feeding on the tall vegetation near our home. When the cow had left I examined the place and found that it had been picking out only the tops of fireweed. You will notice that twigs or plants nipped by hoofed animals with their blunt dental equipment do not show the neat, sharp-cut edge left by the chisel teeth of rodents and rabbits. For example, compare figures 152, b, and 15, d.

In hunting journals we have read much about moose calling, and the birch-bark horns devised for that purpose. There has been much misunderstanding about this. The bull moose in rut produces a low sound, a brief grunt—*Mm-uh!* It is the cow moose, when in a rutting state, that gives out a series of loud sounds,

Figure 150 (opposite)

a. Pellet form, with concave end (Alaska).
b. Double pellet (British Columbia).
c. Long and narrow scats, irregular in shape (Wyoming, 1951).
d. Pellets of irregular shape (Alaska, 1951).
e. From moose calf (Wyoming, December 25, 1950).
f. Large pellets (Minnesota).
g. From young cow moose (January 21, 1952).
h. Soft pellets, some joined, from large bull (Wyoming, August 15, 1929).
i. Pellets coalesced in soft mass, result of succulent feed (Wyoming, August 1951). Scats in h and i are only a small portion of the total sample.

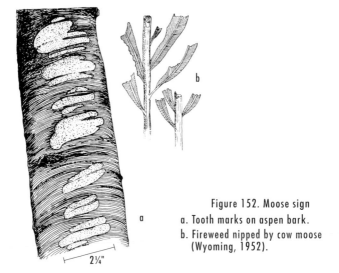

Figure 152. Moose sign
a. Tooth marks on aspen bark.
b. Fireweed nipped by cow moose (Wyoming, 1952).

which may be described as bawlings of a sort. On occasion I have seen a pair together, at a moderate distance, when the female gave out a distinct call, *Uh-u-ow-wa*, and I could see the bull raise its muzzle slightly and open its lips a little. I knew he was calling, but the grunt was so low that it failed to reach my ears.

The cow and calf will call to each other, but the sounds are low. The bleatlike call of the moose calf is in a lower tone than the squeal-like call of the elk calf.

CARIBOU AND REINDEER

The northernmost member of our deer family, the caribou (*Rangifer tarandus*), occupies the circumpolar regions of North America. Caribou now share their niche with their slightly smaller cousin, a subspecies known in the Old World as reindeer, which were brought over to Alaska from Siberia and Europe for domestication. They are the same species, but the domesticated populations are often referred to as reindeer, while their native counterparts are called caribou.

In America the caribou occupy most of Alaska, though not continuously, and the northern tundra and the boreal forests of Canada; formerly they ranged down into Maine, northern Minnesota, and northern Idaho.

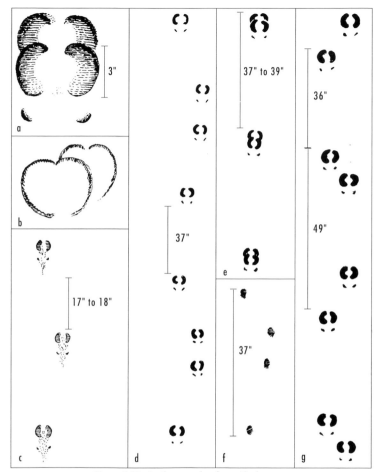

Figure 153. Caribou tracks, from Alaska

a. On wet sand; hoof marks sometimes nearly 5 in. long.
b. On a layer of snow on ice, only the hoof edges making a mark.
c. Walking in shallow snow.
d. Slow gallop. Another trail showed 5-ft. leaps; leaps would be greater with speeding.
e. Trotting, on wet sand.
f. Gallop of a young calf, 20-in. leaps.
g. Trotting in sand, the hind legs straddling the front legs so that the hind tracks register beyond the front tracks on a given side of the body.

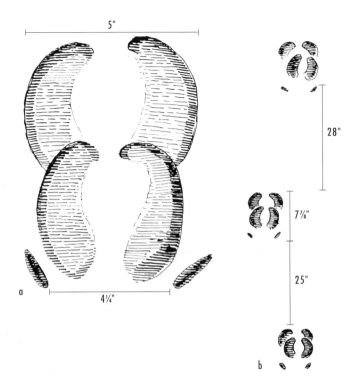

Figure 154. Caribou tracks in mud, walking gait, from Alaska

a. Smaller hind track slightly overlapping the larger front track; dewclaw marks go with the front track.

b. Walking trail pattern; straddle is 9¾ in.

Caribou have the most distinctive track of all the deer, for their hooves are rounded and tend to spread, the snowshoe-like effect thus helping somewhat to support the animal in snow. A track on a thin film of snow, as shown in b of figure 153, is almost circular. Caribou tracks cannot be confused with moose tracks.

There is an interesting characteristic of the caribou foot that is often revealed in the tracks. In the front feet there is a considerable gap between the hooves and the dewclaws, or pasterns. Moreover, the front hooves tend to extend out forward of the vertical line of the front leg, and the dewclaws tend to come low to

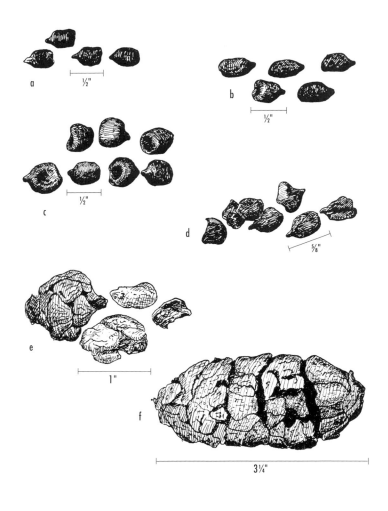

Figure 155. Caribou droppings, about ⅔ natural size (Alaska)
a. From a young female (May 12, 1950).
b and c. Other typical samples (May 10, 1940).
d. Soft pellet type, from female (May 12, 1950).
e. Pellets partly coalesced.
f. Soft type. Scats in e and f both result from green food.

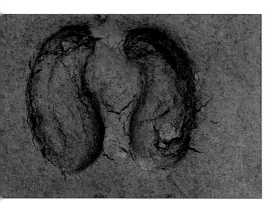

A clean front track of a caribou in Denali National Park, Alaska.

the ground as the animal steps. The hind foot, on the other hand, is more compact, with the dewclaws closer to the hooves, and is held higher from the ground. Hence, in some tracks in mud or shallow snow, when the hind foot partly covers the front track, only the front dewclaw prints show, and these may be far enough back to be associated with the hind track. This is shown in figure 154. Measurements are provided here for caribou, but do NOT include the dewclaws. Front tracks measure 3¼–5 inches long and 4–6 inches wide. Hind tracks are 3–4½ inches long and 3⅝–4¾ inches wide.

It has been observed also that with greater traveling speed the *dewclaw* prints tend to be at right angles to the line of travel, being more diagonal at the slower speeds.

Caribou droppings are not as distinctive as the tracks, and more nearly resemble those of other deer. The coalesced type resulting from succulent feed resembles the comparable types of mountain sheep and some of the other hoofed animals (see figure 155, e and f).

Traveling bands of hoofed animals generally produce vocal sounds as the female and young call to each other. We are all familiar with the baaing of sheep and the bawling of cattle under such circumstances. The caribou vocabulary is a short grunting sound, a low-pitched *a-a-w, a-a-w.*

The caribou also will thresh bushes or young trees with their antlers to remove the velvet and to scent mark. In winter you will find big pits in the snow where they have pawed down with their round flat hooves to find the lichens underneath.

Whether it be in the Canadian northland or the mountains of Alaska, caribou sign is exciting, because it quickens the anticipation of possibly seeing the picturesque animal itself.

PRONGHORN OR ANTELOPE (ORDER ARTIODACTYLA, FAMILY ANTILOCAPRIDAE)

The technical name, *Antilocapra americana,* suggests the unique character of this animal. The latter part of the genus name, *capra,* suggests goat, and so it was called by Lewis and Clark. The pronghorn is a western plains animal, ranging from southern Saskatchewan south into northern Mexico; it is found in Montana, western North and South Dakota, western Nebraska, Wyoming, Idaho, Utah, Nevada, Colorado, eastern Oregon, and parts of California, Arizona, New Mexico, and western Oklahoma and Texas.

This graceful animal is built for speed and loves to exercise this gift. Often I have seen a band take off on a swift race just for the sheer joy of running. Their running gait is smooth and level, with hardly any rise in the air at each bound.

Donald D. McLean of California recorded a jump of at least 27 feet: "I measured two such jumps in Lassen County where they cleared an open mud-bottomed cut." Other trails showed average leaps of 14 feet at high speed.

In calculating the 14-foot leap, Mr. McLean measured from right hind foot to right hind foot. Dr. Morris F. Skinner, who studied running pronghorn in Nebraska, found that the spread of the four feet when landing was 9 feet, with 6 to 7 feet to the next group of tracks—which comes pretty close to the same result.

Dr. Skinner followed a mature buck in Nebraska with a car at 40 to 44 miles per hour. A doe ran 38 to 42 miles per hour. At first the doe held her head high, then laid back her ears and stretched her head and neck forward as she went into high speed.

The pronghorn, or antelope, has no dewclaws, and therefore

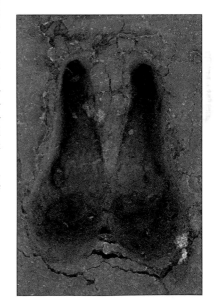

The front track of a pronghorn in southern Colorado.

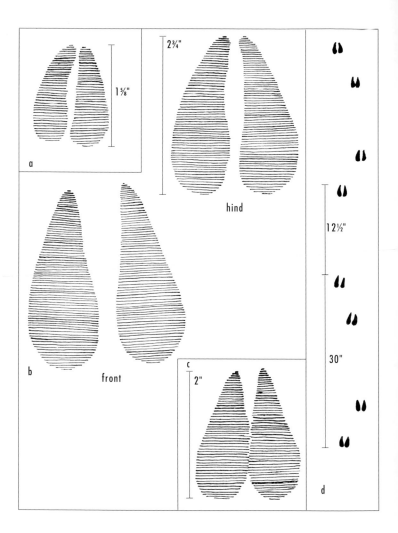

Figure 156. Tracks of pronghorn

a. Fawn track, about ⅔ natural size (Yellowstone National Park, July 16, 1929).

b. Adult tracks, in mud, ⅔ natural size. Front track, 3¼ in. long; hind, 2¾ in. (Nevada).

c. Young pronghorn track, in mud (Nevada).

d. Adult track pattern, a lope (Nevada).

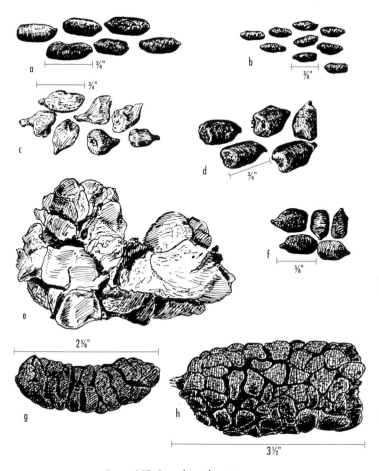

Figure 157. Pronghorn droppings,
all except e, g, and h about ⅔ natural size (Nevada)

a. Pellet type, from adult (August 15, 1939).
b. Pellets of fawn (July 25, 1940).
c. Semisoft pellets, from adult (September 15, 1939).
d. Hard pellet type, from adult (August 28, 1940).
e. Soft mass type (August 16, 1939).
f. Pellet type, from adult male (August 18, 1939).
g. Soft type, from young animal (August 18, 1939).
h. Soft type, with pellet structure apparent.

has the most streamlined foot of all, with the exception of the single-toed horse.

Pronghorn tracks are much like those of deer, but the hind border is usually broader, and the outside edges are often distinctly concave. However, some tracks are difficult to distinguish. Figure 156 presents the general shape. In some areas mule deer share some of the range with pronghorn, and their tracks may be confused. Front tracks of pronghorn measure 2⅛–3½ inches long and 1½–2¼ inches wide. Hind tracks are 2¼–3¼ inches long and 1½–2⅛ inches wide.

In many respects pronghorn droppings resemble those of deer and mountain sheep. The summer droppings tend to form an elongated narrow mass, very similar to certain scat types of the mountain sheep. (Compare, for example, figures 157 and 165.) On some occasions, a first hasty glance at such a sample in the sagebrush country reminded me of the coyote scat.

The pronghorn has an interesting habit of scraping the ground with a hoof to mark its territory, then depositing the dropping or urine on the bare spot of ground. This is the reverse of the bobcat's habit of *covering* its droppings by scratching dirt and debris toward it. Thus the two types of scratch signs can be distinguished, even without reference to the distinctive scats.

When startled or worried the pronghorn produces a loud whistling sound similar to that made by the white-tailed deer, but it seems to me more musical, with almost a vocal quality—a fairly high-pitched *hieu-u-u-u!* The doe-to-fawn calls are a lower bleat. When annoyed or curious the pronghorn will stamp a foot as the deer do.

BISON, CATTLE, GOATS, MUSKOX, AND SHEEP (ORDER ARTIODACTYLA, FAMILY BOVIDAE)

AMERICAN BISON OR BUFFALO

The American bison, or buffalo, *Bos bison*, can now only be found in a few places: in northern Alberta, the Alaska Range south of Fairbanks, on the Bison Range south of Glacier National Park, in the Yellowstone and Wind Cave National Parks, in House Rock Valley in Colorado, on the Wichita Mountains Wildlife Refuge in Oklahoma, and several other places.

The tracks are very similar to those of domestic cattle. On hard ground, where only the outer rim of the hooves leave a clear mark, the cloven-hoof effect is not prominent; the first impression may be that of a horse track. The droppings, too, are very

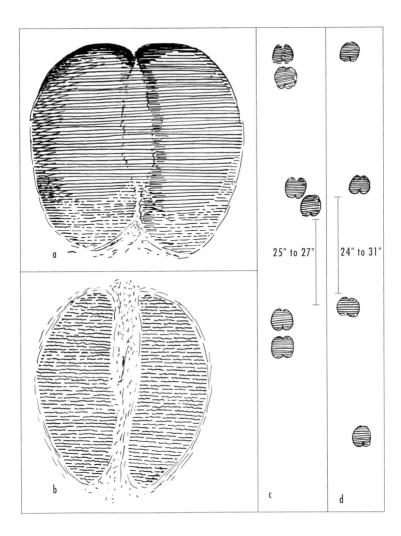

Figure 158. Bison tracks

a. Typical track in mud, 5 × 5 in. (Yellowstone National Park).

b. Restored track of extinct superbison from Alaska (from plaster cast of left front foot of a young Pleistocene specimen obtained by Otto W. Geist).

c and d. Two track pattern variations in bison walking. Forefoot tracks are slightly larger than those of the hind foot (Yellowstone National Park, 1929).

similar to those of cattle, both in the soft formless and the harder layered types. Figure 158, b, shows a "restored" track of a young Alaskan Pleistocene superbison, obtained from a plaster cast made from the actual foot of a specimen. Of course, the hooves of dead beasts shrink considerably; an actual track would be larger, but it closely resembles the track of a modern bison.

Front tracks measure 4½–6½ inches long and 4½–6 inches wide. Hind tracks are 4½–6 inches long and 4–5½ inches wide.

In Hayden Valley of Yellowstone, I came upon a pine growth

Figure 159. Bison sign

a. Buffalo wallow on the plains, where the animals roll and produce a slightly depressed bare place in the dirt.

b. Chips or droppings of the soft type, when diet is succulent, drying into a flat mass; diam. about 12 in.

c. Chips of the harder, layered type, with drier feed.

Buffalo tracks in Yellowstone National Park.

where bison had rubbed and horned trees extensively, so that most of the trunks had a light-colored, worn ring. There were the characteristic long, brown, somewhat kinky hairs of bison clinging to the bark. In Oklahoma I have found isolated trees worn so smooth in this way that they were finally killed. Bison will also rub against big boulders, telephone poles, or any other convenient scratching place. You will find the trampled ground at the base, and usually a few bison hairs.

The buffalo wallow has always been a feature of the plains (figure 159, a). Like horses, these shaggy animals are fond of rolling in dust, but I have never seen one roll entirely over as a horse does. I suppose the hump is in the way. It is reported that in early days bison used to tear up a wet, soggy place with the horns, then roll about until their bodies became caked with mud. This resembles the wallowing of bull elk in the rutting season, and may have had the same significance.

DOMESTIC CATTLE

In many parts of the West domestic cattle occupy the same range as elk and other large mammals, and in some instances their tracks may be confusing. Domestic cattle tracks resemble those of bison and muskox. The track of a large calf or a yearling is easily confused with adult elk tracks. The mature domestic cow or bull track, however, is larger and more blocky than that of elk.

Cattle droppings also are similar to those of bison, with the same variations. It should be noted that certain elk droppings resulting from succulent feed in summer are very similar to those of cattle at that season. It would be difficult to distinguish them, ex-

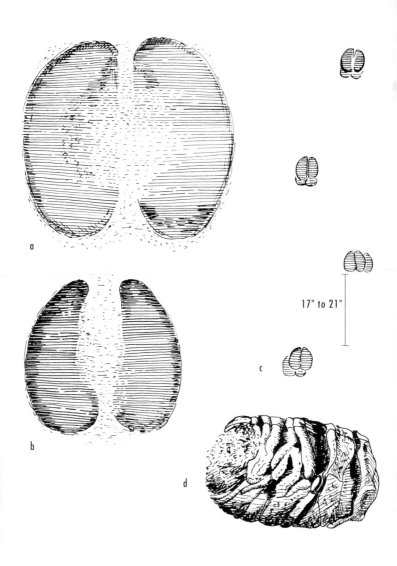

Figure 160. Domestic cattle sign, from Wyoming
a. Adult Hereford track, in mud; length 4¼–4¾ in.
b. Large calf track, length 3 in.
c. Hereford cow track pattern, 17- to 21-in. steps.
d. Dropping, firm type, 6½ in. long.

cept that on the average the size, or quantity, of an elk dropping is much smaller than that of domestic cattle. Compare figure 160 with figures 131 and 139.

MOUNTAIN GOAT

The ace mountaineer of our big game animals is the mountain goat, *Oreamnos americanus,* which occupies high mountain crags from southeastern Alaska south and east through British Columbia into the Cascade Range of Washington and the Rocky Mountains in Idaho, western Montana, the Yellowstone area, and parts of Colorado. This animal has also become established in the Black Hills of South Dakota.

Tracks of the mountain goat resemble those of mountain sheep in that the toes tend to spread so as to present a somewhat square track form, as shown in figure 161. Front tracks measure 2½–3½ inches long and 2½–3½ inches wide. Hind tracks are 2½–3¼ inches long and 2¼–3 inches wide.

The droppings are easily confused with those of deer and sheep, as comparison with the illustrations for those animals will reveal, though they tend to be somewhat smaller. It is instructive to note the transition in form of the droppings, from the mass formed by succulent vegetation (figure 161, i), through the clustered pellets in h, the soft pellets of d and e, to the hard winter pellets in f and g when the animals are on dry feed. Note in g that the large samples appear to be made of two pellets combined.

As with other mountain dwellers, the mountain goat makes beds on rocky ledges, where the dusty beds and droppings can be found. Like mountain sheep, it also takes refuge in caves.

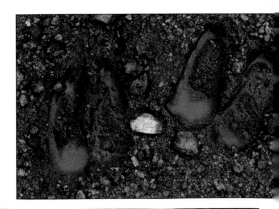

*Front and hind tracks
of a mountain goat in
central Colorado.*

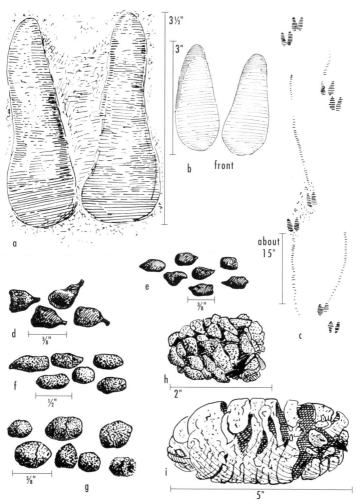

Figure 161. Mountain goat tracks and droppings

a. Typical track in snow.
b. Front track in wet sand (Colorado).
c. Trail pattern, walking, in light snow.
d. Droppings, late summer or early fall.
e. Droppings of kid, late summer.
f and g. Winter droppings.

h. Summer droppings, late enough to form pellets, which, however, combine in a cluster.

i. Summer dropping when animal is on succulent feed.

There are domestic goats of many breeds, and their tracks will vary accordingly in size. In general, their tracks and droppings are nearest in characteristics to those of the native mountain goat, as shown by comparison of figures 161 and 162. Domestic goats do not occupy the present high ranges of the mountain goats in the Northwest, so there should be no confusion in the localities occupied by the latter. In the southwestern United States and in Mexico, however, domestic goats often range in the high country occupied by mountain sheep and deer. Consequently, in those areas the tracks encountered must be scrutinized carefully. The tracks shown here, especially in figures 161, 162, and 166, should help to distinguish this domestic animal's track from those of the wild ones.

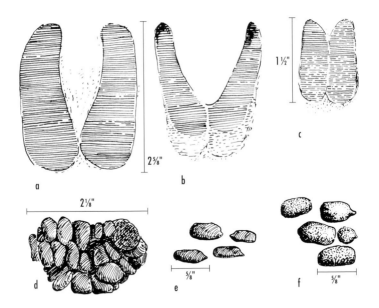

Figure 162. Domestic goat tracks and scats
a and b. Tracks in mud (Kanaga Island, Alaska, 1937).
c. Track of kid, in mud (Kanaga Island).
d. Summer dropping (Kanaga Island).
e. Winter droppings (Wyoming).
f. Winter droppings (Texas).

The muskox, *Ovibos moschatus,* has chosen the farthest-north lands to live in, and few of us have the opportunity to find the animal at home. But in 1936 I landed on Nunivak Island, in the Bering Sea, where some muskoxen had been released. There among the sand dunes was an old bull with all the appearance of

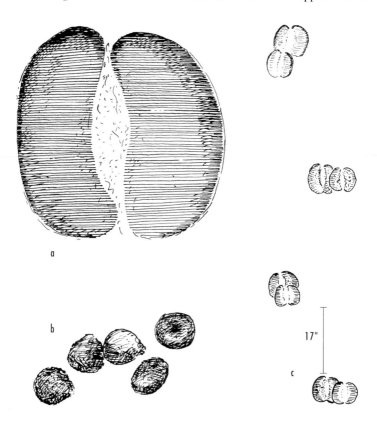

Figure 163. Muskox tracks and droppings, from Nunivak Island, Alaska
a. Track in mud, about 5 in. length and width.
b. Droppings in pellet form.
c. Trail in sand. Straddle is about 10 in. Front-foot print is 5 × 5 in.; hind-foot print, 4 × 4 in.

belligerence as I approached for pictures. There I found its tracks, so much resembling the tracks of bison and domestic cattle.

Muskoxen are strange creatures that are content to live on the Arctic tundra of northern Canada, in Greenland, and even on some of the storm-swept islands of the Arctic Archipelago, where their long warm coats serve them well. They have a unique habit. When a band of muskoxen is approached by wolves, or by human hunters, all the adults form a circle, with their horns facing out against the enemy, and the calves are gathered in the center behind this array of natural weapons. In earlier days, whenever an expedition desired to capture a muskox calf for a zoo, it was necessary to shoot down the defensive circle of parent animals before the calves could be taken.

The tracks shown in figure 163 from Nunivak Island are much like those of cattle, but you will find no cattle on natural muskox range.

ALL'S AND BIGHORN SHEEP

The mountain sheep, of the genus *Ovis,* inhabit the backbone of our continent, including the mountain ranges of Alaska and Yukon Territory down through the Canadian Rockies in British Columbia and Alberta, through Idaho and Montana, Wyoming, Nevada, Utah, Colorado, California, Arizona, New Mexico, western Texas, northern Mexico, and Baja California.

Mountain sheep tracks also resemble those of deer. Generally, however, the hooves have straighter edges and the tracks do not so often take the traditional heart shape. In other words, they tend to be more blocky, somewhat less pointed. Nevertheless, it

Bighorn sheep tracks in a mud puddle in central Colorado.

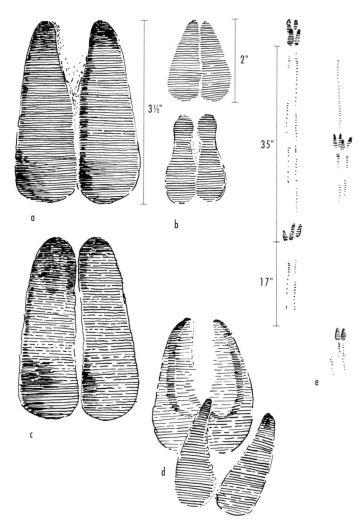

Figure 164. Mountain sheep tracks
a. Track of bighorn in mud (Wyoming).
b. Bighorn lamb tracks in mud (Wyoming).
c. Track of desert bighorn in mud (Death Valley, California).
d. Tracks of Dall's sheep (Alaska).
e. Walking pattern of Dall's sheep in snow (Alaska).

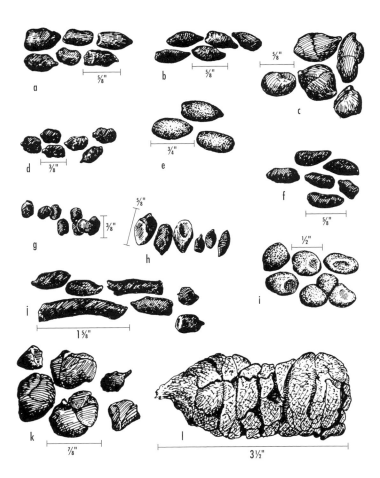

Figure 165. Mountain sheep droppings, about ⅔ natural size

a. Dall's sheep ram (Alaska).
b. Dall's sheep (Alaska).
c. Desert bighorn sheep (Texas).
d. Yearling Dall's sheep (Alaska).
e and f. Desert bighorn sheep (Texas).
g. Dall's sheep lamb (Alaska).
h. Bighorn ewe and lamb (Wind River Mountains, Wyoming).
i. Bighorn clay droppings, from mineral lick (Jackson Hole, Wyoming).
j. Bighorn, on soft feed (Wyoming).
k. Bighorn, soft pellet type (Wyoming).
l. Bighorn scat, result of succulent feed (Wyoming).

cannot be denied that certain samples are confusing. Front tracks measure 2⅛–3⅜ inches long and 1½–3 inches wide. Hind tracks are 2¹⁄₁₆–3¼ inches long and 1½–2⅜ inches wide. The tracks in figure 164 were chosen as the most undeerlike.

As for the droppings, note the variety in figure 165 and also the variety shown for the different kinds of deer. Tracking is not easy, and one must take into account the location of signs, whether among mountain sheep ledges or in deer forest.

The bighorn sheep make beds which may be looked for on jutting points above cliffs, wherever the animal has a good view. The animal scrapes at the dirt with a hoof to make a slight smooth hollow, then lies down. These beds are used often, and in areas long occupied by mountain sheep there will be a considerable accumulation of dung.

The voice of the mountain sheep is the traditional *baa* like that of domestic sheep. However, bands of mountain sheep are generally rather silent, in contrast to the noisy domestic sheep.

My brother and I once took advantage of this call in photographing the Dall's, or white, sheep in Alaska. I was on a ridge, waiting to photograph a group of sheep coming up toward me. They were down the slope, out of my view, and I was uncertain as to just where to expect them to appear on the ridge. My brother was at one side on the slope, where he could see me and the sheep, and, imitating the voice of the sheep so as not to alarm them too much, he guided my actions on top: "*Baa-a-a-a*—they're moving to your right —*baa-a-a-a*—now they're coming straight up."

Thus guided, I found myself in camera range when they came out on top.

If you are fortunate, you may have the experience of witnessing an interesting incident in mountain sheep activity. In the autumn, if you are in mountain sheep country, you may hear a loud *whack* as if someone had slammed two stout boards together, or had banged shut a car door. Follow the sound, and you may come upon two jealous rams fighting by simply banging their heads together.

DOMESTIC SHEEP

Tracks and droppings of domestic sheep may be confused with those of deer, mountain sheep, and goats (compare figures 140–47, 161–62, and 164–66). In the western plains and mountains domestic sheep travel in large bands. If you encounter a large mass of trails leading across the country, with hoofprints about as shown in figure 166, you may be pretty sure that domestic sheep have traveled there. Deer are solitary or in small groups. Mountain sheep and domestic goats also are pretty well scattered, or in groups of a few individuals.

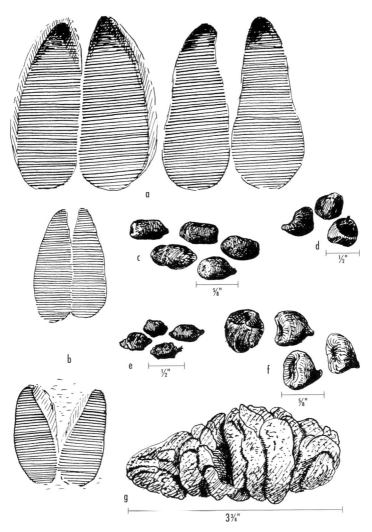

Figure 166. Domestic sheep sign

a. Two typical tracks of ewes, in mud; about ⅔ natural size (Utah).
b. Tracks of lambs, in mud; about ⅔ natural size (Utah).
c. Droppings of ewes, hard type (Wyoming).
d, e, f. Semisoft pellet types (from Texas, California, and Utah, respectively).
g. Dropping characteristic of succulent feed (Utah).

With reference to individual trails, the deer is longer-legged than the domestic sheep and takes longer steps. That is one clue.

It is worth while to learn, if possible, whether or not there are domestic sheep ranging in a given locality. Their known absence simplifies the problem of identification of tracks and droppings. Field biologists often take advantage of such information in the hope of narrowing the search to only a few possibilities. On one occasion I found evidence in a part of the Teton mountain range, Wyoming, of occupancy by either domestic or mountain sheep. It was important to know which, in order to appraise a wildlife range problem there. I was unable to identify the abundant droppings that I found on a high mountain slope. It was only by finding the old remains of a dead mountain sheep and obtaining information on the ranges allotted to the grazing of domestic sheep that I could be reasonably sure the area in question had been occupied by mountain sheep.

Figure 166 shows what differences may exist in the tracks and droppings. It also, I fear, reveals confusing similarities.

BAIRD'S TAPIR (ORDER PERISSODACTYLA, FAMILY TAPIRIDAE)

Tapirs are relatives of the horse and rhinoceros, and are found in the Malayan region and in the American tropics. The species that inhabits Central America and part of Mexico is *Tapirus bairdii*, an animal somewhat smaller than the other species.

In the Panama Canal Zone I found the distinctive three-hoofed tracks in the mud where the animal had crossed little streams. Sometimes the well-marked footprints followed the man-made

Look carefully for the complete front and hind tracks of a tapir, intermingled with the tracks of a jaguar on a beach in Costa Rica; photo by Marcelo Aranda.

jungle trails. The tapir may also have its own well-worn trail, resembling a cattle trail except for the three-hoofed tracks in the mud. Tapirs readily take to the water and are known to wallow in shallow pools.

In figure 167, a, the front track shows the fourth toe on the side, but this may not be easily seen in many of the tracks you find, especially when the hind foot steps into the front track. The straddle of the track pattern, as I found it, was slightly over 10 inches.

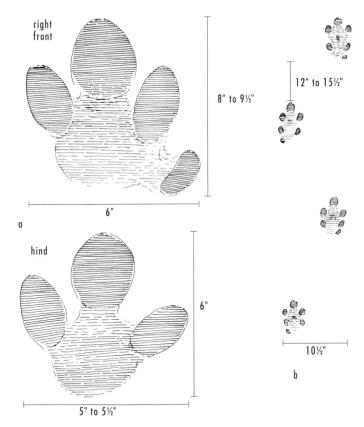

right front

8" to 9½"

6"

a

hind

6"

5" to 5½"

12" to 15½"

10½"

b

Figure 167. Tracks of Baird's tapir in mud (Panama Canal Zone)
a. Front and hind tracks.
b. Walking trail pattern. The hind foot appears to slightly overstep the front track.

BIRDS

BIRDS, WITH THEIR flight, songs, and bright colors, can be found and recognized without undue reliance on tracks and other sign. Yet a knowledge of a few of these signs is helpful, and in some respects necessary for certain studies. In the following

Left and right tracks of a roadrunner in Death Valley National Park, California.

The walking trail of a mourning dove in Texas's coastal dunes.

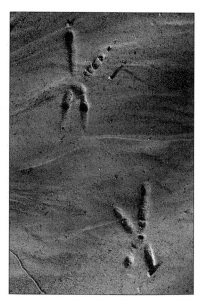

pages are given a few tracks of representative bird groups, those most likely to be encountered on tracking expeditions, though one, the condor, can no longer be considered a likely find.

In general perching birds—those, like sparrows and a great many others, that spend much of their time in the trees—tend to hop when on the ground, so their tracks appear paired. All game birds and those that spend much time on the ground, such as the raven, magpie, robin, and pheasant, walk or run. Thus you will find the first group mostly hopping, leaving parallel tracks, and the second group leaving mostly alternating tracks. There are of course some exceptions—you will find some robin tracks paired, for example. Also, the great horned owl, which lives mostly in trees, walks when on the ground. (See figures 168–76.)

Bird droppings, a number of which are illustrated here (figures 177–79), are readily distinguished from those of mammals, yet some are not easily distinguished from each other. As with mammals, in some cases the hard dropping denotes dry food, such as twigs and buds for grouse, while the soft, semiliquid ones result

A bald eagle's trail along an Alaska river.

The walking trail of a burrowing owl in dunes of the Mojave Desert in southern California.

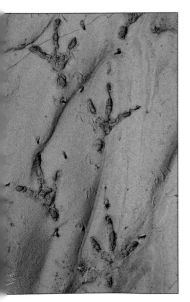

from succulent food. Bird droppings tend to be coated, at least at one end, with white calcareous material.

The pellets, or castings (figures 180–82), that are habitually regurgitated by some birds are at first glance similar to the scats of certain carnivores, but bird castings consist of pure feathers, fur, and bones, free from the digestive residue mingled with such remains in carnivore droppings. This residue in scats appears as crumbling, dirtlike material that serves as a matrix for the included hair and bones.

Bird pellets are extremely useful in studying the diet of certain birds. However, a study of the series illustrated here reveals so much variation in size and shape for each bird, depending on the kind of food eaten, that it is impracticable to identify the individual pellets without reference to a known nest or perch. Note, for example, the extreme variations in the pellets of the great horned owl (figure 182, c); note also the general similarity between those of this owl and those of the golden eagle. I have found it necessary to determine in some other way which bird uses a particular perch, or nest, where pellets are collected for study.

In spite of all these limitations, much natural history is revealed by a knowledge of bird sign. An accumulation of character-

Lower left: 3 sanderling pellets; top center: 1 belted kingfisher pellet; right: 2 Heermann's gull pellets (a small gull species). All were collected from Pescadero State Beach and Marsh, San Mateo County, California. (The penny is ³⁄₄ inch in diameter.)

istic droppings (figure 178, e) on and beside a log reveals the drumming log of a ruffed grouse. Specific tracks concentrated about a bit of carrion in the woods disclose to you that the sharp-eyed ravens and magpies have found it. A mass of white guano streaks in vertical smears on a cliff face will point to the location of the aerie of certain hawks, falcons, owls, or raven. A splashing of similar white stains among the green vegetation in the woods should cause you to look overhead for a possible nest. There is satisfaction in reading the story of streamside loiterers by the tracks of goose, duck, or gull, especially if droppings or an occasional feather help you to make sure of the tracks. It is well to note whether it was a crane, with short hind toe, or a heron, with the long hind toe, which had walked in the soft mud.

You may be puzzled, as I was once, to know whether a crane or a wild turkey had walked on the shore of a pond. The crane foot is trim, with a fine "fingerprint" pattern. The turkey foot is coarser, with rougher undersurface.

Thus, it seems worthwhile to give some passing attention to birds in this guide devoted principally to mammals.

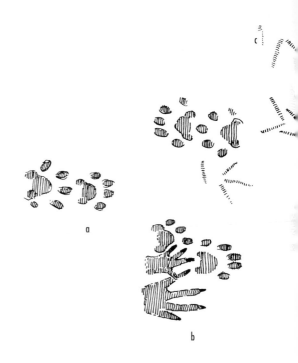

a

b

c

Figure 168. Riverbank record

Certain animals are wont to travel along riverbanks. Here a mink (a) has happened along, in both directions. A muskrat (b) passed to the right. Spotted sandpipers (c) left their tracks. A Canada goose (d) has walked over the tracks of the sandpiper and muskrat. Part of the web shows in only one track, the upper one. And, in the upper right, are some marks that may possibly represent imperfect tracks of a toad (e). Here, then, is a traffic record emphasizing the habits of a group of animals that have mingled their footprints in a place each one considers home ground. Notice that the goose footprint does not always show the web, and that the prints of the sandpiper do not always show every toe.

e

d

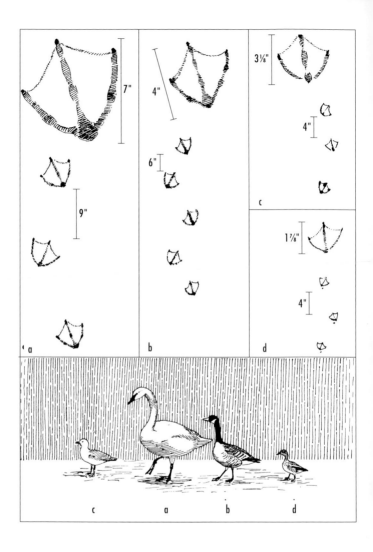

Figure 169. Waterfowl tracks
a. Trumpeter swan, in mud (Wyoming).
b. Canada goose, in mud (Wyoming).
c. Glaucous-winged gull (Washington).
d. Green-winged teal (Aleutians).

Figure 170. Tracks of grouse

a. Sage grouse, in mud.
b. Blue grouse, in snow.
c. Ruffed grouse, in mud.
d. Male ring-necked pheasant, in mud.
e. Rock ptarmigan: left, in mud; right, in snow, when toe feathers are well grown.

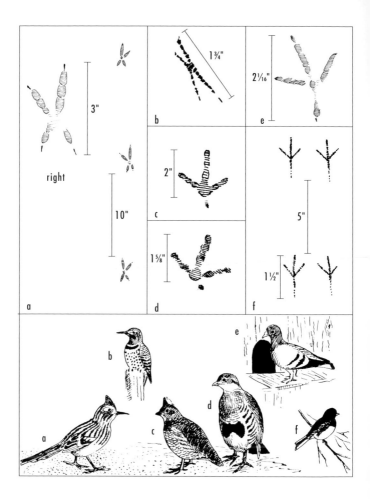

Figure 171. Miscellaneous bird tracks

a. Greater roadrunner, in mud (California).
b. Common flicker, in mud.
c. Scaled quail, in dust (Texas).
d. Gray partridge, in dust (Nevada).
e. Domestic pigeon, in mud (Massachusetts).
f. Dark-eyed junco, in snow.

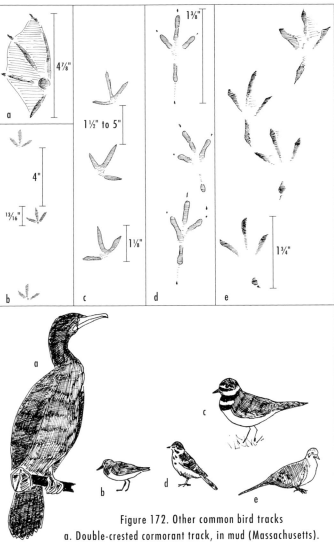

Figure 172. Other common bird tracks
a. Double-crested cormorant track, in mud (Massachusetts).
b. Trail of sanderling, in moist sand (Texas).
c. Killdeer trail, in mud (New Hampshire).
d. American pipit walking in mud (Massachusetts).
e. Mourning dove walking in sticky, deep mud (Massachusetts).

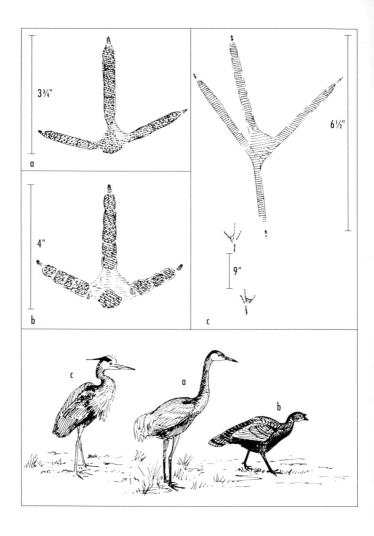

Figure 173. Tracks of crane, turkey, and heron

a. Sandhill crane, in mud (Oklahoma).

b. Wild turkey, in mud (Oklahoma).

c. Great blue heron, right foot, in mud. Below is the stride, almost in a straight line (Yellowstone National Park).

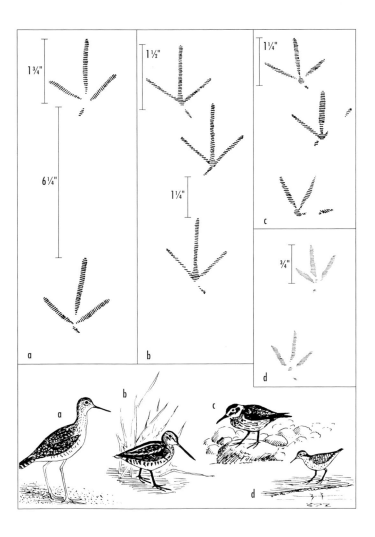

Figure 174. Some shorebird tracks
a. Greater yellowlegs, in mud.
b. Common snipe, in mud.
c. Rock sandpiper.
d. Spotted sandpiper, in mud.

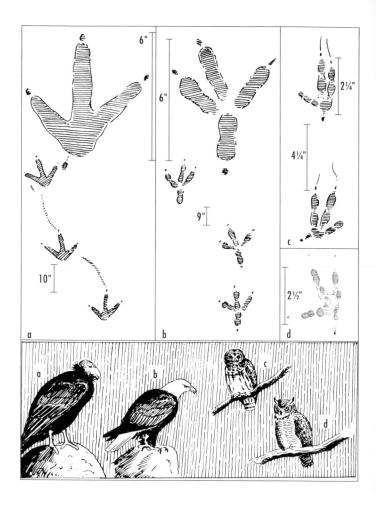

Figure 175. Tracks of some birds of prey

a. California condor, in snow (National Zoological Park, Washington, D.C., 1929).

b. Bald eagle, on sand (Aleutians).

c. Barred owl, in snow (Washington, D.C.).

d. Great horned owl, in mud (Death Valley National Park).

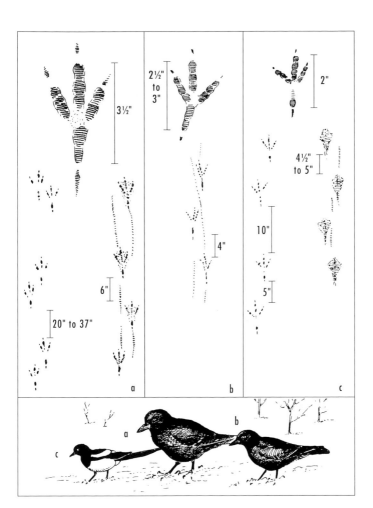

Figure 176. Tracks of raven, crow, and magpie

a. Common raven track, in sand (Aleutians): left tracks, hopping to take flight; right, walking in snow (both from Wyoming).

b. American crow track, in mud (Oklahoma); below, walking gait.

c. Black-billed magpie track in snow (Wyoming): left, skipping trail; right, walking, in deeper snow.

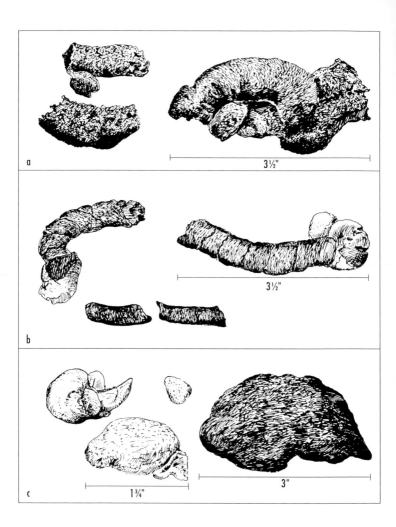

Figure 177. Bird droppings, about ⅔ natural size

a. Trumpeter swan (Yellowstone National Park).

b. Canada goose: upper left, Semichi Island, Aleutians; upper right, Canada; lower, cackling Canada goose from Buldir Island, Aleutians.

c. Great blue heron. This material is puzzling, for the largest specimen appears to be a disgorged pellet. The diet in this case is field mice (Wyoming).

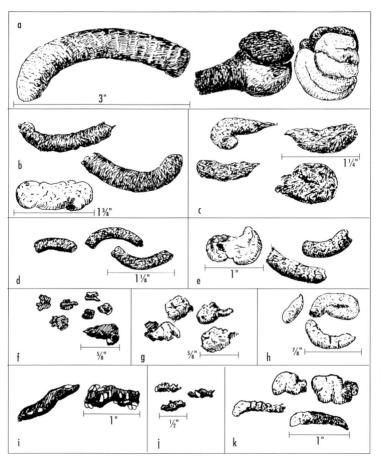

Figure 178. Bird droppings, about ⅔ natural size

a. Wild turkey, the scat of a male gobbler (left, Arizona) and (right) typical scats of females (Oklahoma).

b. Blue grouse: upper winter sample is old, dried out to smaller diameter; middle sample is fresh, somewhat larger; lower left is a summer sample, softer (all Wyoming).

c. Sage grouse.

d. Spruce grouse (Minnesota).

e. Ruffed grouse: left sample is from summer food.

f. Harlequin quail (Texas).

g. Lesser prairie-chicken (Oklahoma).

h. Sharp-tailed grouse (North Dakota).

i. Bohemian waxwing, on berry diet.

j. Townsend's solitaire, on berry diet.

k. Chukar (Wyoming).

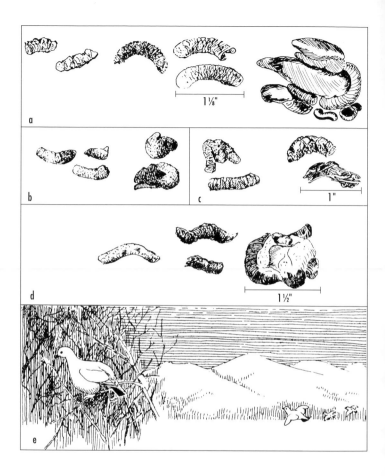

Figure 179. Bird droppings, about ⅔ natural size

a. Rock ptarmigan (Alaska): right, a soft-type summer specimen; two center ones (Denali National Park, Alaska) and those at left (Aleutians) show variations of the hard type.

b. White-tailed ptarmigan (Colorado).

c. Rock ptarmigan: left, Denali National Park; right, Alaska Peninsula.

d. Common raven droppings: right, consisting of coiled material partly coated with white; left, shorter lengths.

e. Ptarmigan in northern landscape. Their winter feeding on buds and twigs produces the woody character of the droppings.

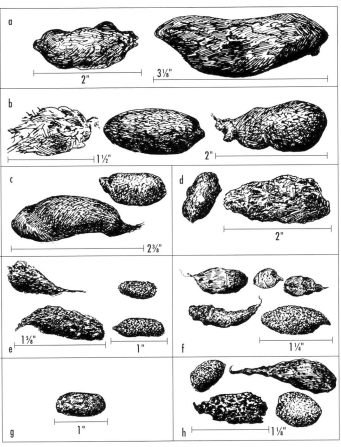

Figure 180. Bird pellets, or castings

a. Ferruginous hawk (Wyoming).
b. Red-tailed hawk: left one contains feathers; the right, ground squirrel fur (Wyoming).

c. Prairie falcon (Wyoming).
d. Swainson's hawk (Nevada).
e. Loggerhead shrike (left, Wyoming; right, Nevada).

f. Sparrow hawk (Nevada).
g. Clark's nutcracker (Wyoming).
h. Black-billed magpie (Wyoming).

Pellets of the gyrfalcon found at a nest on Savage River, Alaska, in 1951 were indistinguishable from those of the prairie falcon (c). That is, they had the same variations in size and shape that you will find at a prairie falcon nest. Both had fed extensively on rodents.

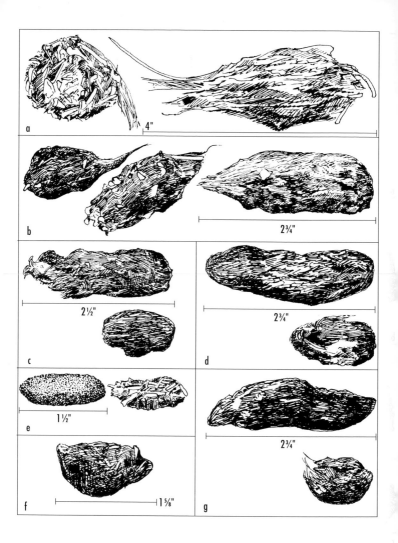

Figure 181. Bird pellets, or castings, about ⅔ natural size

a. Glaucous-winged gull: left, bones and sea urchin fragments; right, feathers (Aleutians).

b. Common raven (Wyoming).

c. Long-eared owl (Wyoming).

d. Short-eared owl (Alaska).

e. Burrowing owl: left, insect remains; right, rodent remains.

f. Great gray owl (Wyoming).

g. Northern goshawk (Wyoming).

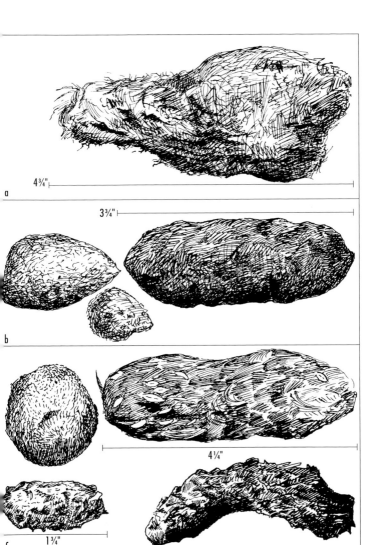

4¾"

a

3¾"

b

4¼"

1¾"

c

Figure 182. Pellets, or castings, of birds of prey, about ⅔ natural size

a. Bald eagle (Aleutians).

b. Golden eagle. These variations could be found at a single nest (from Nevada, Wyoming, and Alaska, respectively).

c. Great horned owl (Wyoming).

AMPHIBIANS AND REPTILES

AMPHIBIANS AND REPTILES too leave signs of their presence. As a matter of fact, in some places when you read the early morning record on a dusty road, you will find toad tracks prominent in the story. You will find, also, that when left to itself and not disturbed, a toad walks and does not hop (see figure 184). However, I have not seen evidence of the frog's walking.

On the desert sands you will find many trails, including those of the lizard, and many small half-moon shaped burrows dug by reptiles of varying size. Tortoises dig large burrows, some of which are later used by burrowing owls. The gopher tortoise of Florida digs a burrow many feet long, with an opening as much as 8 by 12 inches.

There are at least three species of tortoises of the genus *Gopherus* which have accustomed themselves to living in dry, or even desert, areas, and to digging burrows in the ground for shelter. The gopher tortoise, *Gopherus polyphemus,* is the species you will find in the Southeast, from southern South Carolina south through central Florida and west along the Gulf Coast to southeastern Texas. This species is colonial, and burrow openings are oval in shape, flattened in the proportions of the tortoise itself, approximately 6 by 9 inches in diameter, sometimes larger, and they often extend underground for distances of more than 20 feet.

The Texas tortoise, *G. berlandieri,* a little smaller in size, lives only in parts of southern Texas and in Mexico.

Still another is called the desert tortoise, *G. agassizi,* which lives in the Southwest: "Deserts of southeastern California, the southern tip of Nevada, the extreme southwestern corner of

Figure 183 (opposite)

a. American toad.
b. Bullfrog.
c. California newt.

d. Lizard sp.
e. Painted turtle.
f. American alligator.

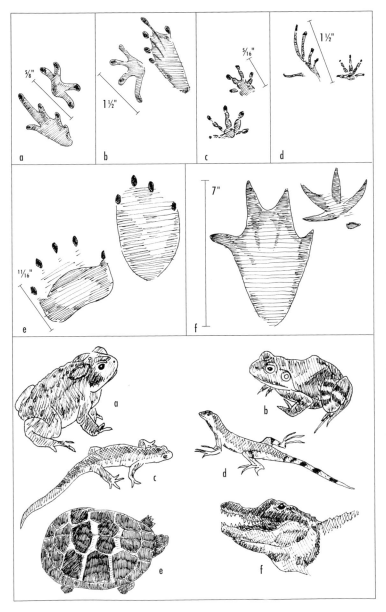

Figure 183. Tracks of reptiles and amphibians

The beautiful trail of a crocodile traveling across a beach in southern Mexico. The straddle of this mature crocodile's trail was nearly 3 ft.

Utah, western and southern Arizona, and Sonora, Mexico" (Clifford H. Pope, *Turtles of the United States and Canada*). The burrows of the desert tortoise are not as deep as those of the eastern species, and apparently these are also more solitary. In fact, some of them find shelter without digging a burrow.

Among the many kinds of burrows you may find within the areas listed here, along the southern belt from South Carolina and Florida to southern California, there are certain oval or somewhat flattened ones that are likely to be those of the gopher tortoise. When you find a burrowing owl at its underground home, or an opossum, rattlesnake, or other animal in its subterranean refuge, the hole it is living in may be one that was originally dug, and that may even be jointly inhabited, by a tortoise.

Reptile droppings are interesting in that, at least among snakes and lizards, you will find a capping of white calcareous deposit at one end, as in the case of birds. Compare figures 185, f, and 187, d, with figures 177, b, 178, a, and 179, d, to mention only a few. Snake scats are often reminiscent of large weasel scats. Toad scats, which do not show a white cap, are often misidentified as small skunk scats.

These few illustrations of amphibian and reptile sign are only a suggestion of what can be found in this particular realm of nature.

On the same beach where I photographed the crocodile trail were numerous trails of hawksbill sea turtles coming ashore to lay eggs. The trail of the ATV belonged to the researcher who moved the eggs to an enclosure to protect them from marauding coatis, raccoons, opossums, skunks, and others.

(Below left) The trail of a Pacific giant salamander in San Mateo County, California. (Below right) The trail of a large snapping turtle in southern New Hampshire. The straddle approached 2 ft.

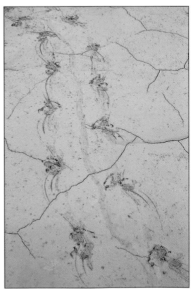

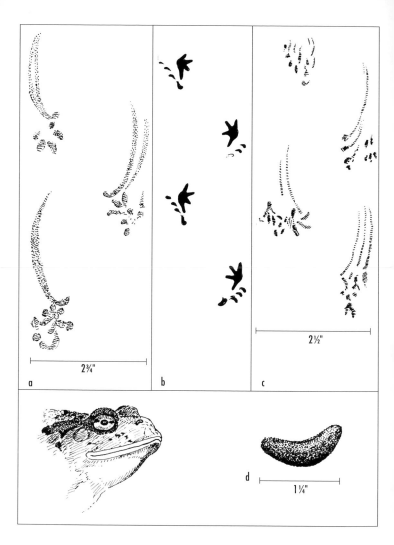

Figure 184. Toad sign

a and b. Toad tracks on a dusty road, walking, with the feet dragging. The exact form of the footprints is hard to determine in dust.

c. Toad footprints in wet mud. Here the toed-in front feet are clearly indicated, as well as the row of toe prints of the hind foot.

d. Toad dropping, about ⅔ natural size.

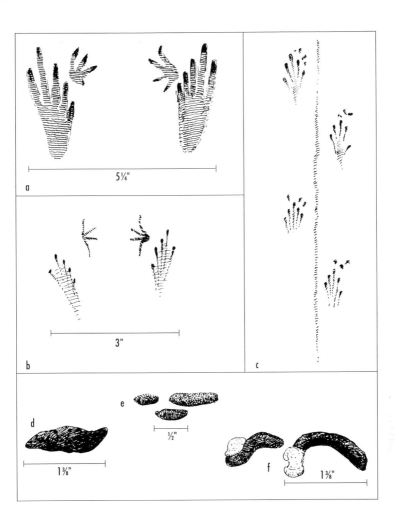

Figure 185. Frog and lizard sign

a. Tracks of bullfrog in mud.
b. Tracks of frog, about size of leopard frog.
c. Lizard track, in sand. The marks of front toes are obscure (Nevada).
d. Dropping of leopard frog (Minnesota).
e. Droppings of small horned lizard.
f. Droppings of collared lizard that had been feeding on insects (Oklahoma).

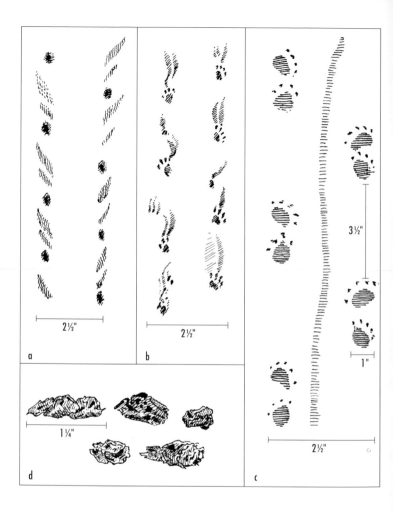

Figure 186. Tracks and scats of turtles, of unknown species (Oklahoma)
a. Tracks in loose sand that rolled into the imprints to obscure their form.
b. On firmer sand, where details are a little more clear.
c. In mud, showing details of feet, and the tail drag.
d. Scats, about ⅔ natural size, black or dark green in color.

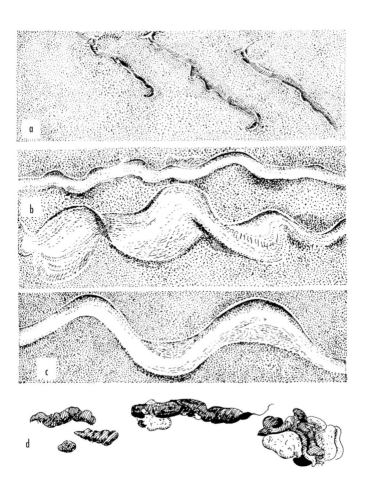

Figure 187. Tracks and scats of snakes

a. Track of sidewinder, *Crotalus cerastes,* in sand; direction of rattlesnake's travel is to the left (from published photograph by Walter Mosauer).

b. Garter snake tracks in dust, moving to right. Above (diam. ⅝–¾ in., moving leisurely; below (diam. ⅝–5 in.), moving fast, with strong side loopings.

c. Track of kingsnake in dust; diam. 1 in.

d. Two samples at right, droppings of hognose snake. At left, droppings of unknown snake species.

INSECTS AND
OTHER INVERTEBRATES

INSECTS LEAVE SIGNS, too, even footprints and droppings. In Florida during the spring of 1952 we found in the sand the raised ridges, the ropelike tubes, that mark the tunnels of the mole cricket (see figure 189, a). These were much like mole runways, though only a scant inch in diameter. There were also little mounds, like miniature molehills, which I learned were the "throw-ups" of certain burrowing beetles of the genus *Geotrupes*. Some mounds were as much as 6 inches in diameter, and when fresh showed a lumpy surface, just as certain fresh molehills do. However, after wind or rain or drying out, they leveled off into a smooth surface. Other beetles that throw up mounds are *Bolboceras* (shown in figure 189, b), *Bolbocerosoma*, and *Phanaeus carnifex*. Some of these mounds are even more inconspicuous. You may also become familiar with the burrowings under the bark of trees, which present characteristic tracery in dead trees when the bark has fallen off.

The trail and signs of leaf-cutter ants in southern Mexico.

As could be expected, the big, bumbling Mormon cricket leaves a distinct trail in dust or sand, as shown in figure 190, a and b. The grasshopper, too, makes a somewhat similar trail when it is not hopping or flying, and of course it comes in many sizes with its many species. One day in sand dune country of the Southwest, I took a walk to see what had been about. There were tracks of lizards, kangaroo rats, and desert foxes—and many little fine traceries that could only be made by insects. There before me was a shiny black beetle making one of those trails (figure 190, c). Notice the similarity of pattern in trails of crickets, grasshoppers, and beetles. Apparently it is the hind leg that trails behind and leaves the backward-pointing mark, and one of the forward pairs, probably the middle pair, that makes the cross mark. You will notice in figure 190, b, which has a more exact cross mark than a, that each print cluster consists of three parts, accounting for the three feet on each side.

Here is a curious problem. Several of us were lounging on the shore of Jackson Lake in Wyoming. In the fine sand were many tracks, including the trail shown on page 360. A deer track? Impossible to imagine an elfin deer that leaves a footprint one third of an inch long! Holding questions and mysteries is of the greatest importance, so I provide one hint and leave you to puzzle over this trail: the "deer tracks" are not the footprints of feet walking, but the tracks of tiny feet digging.

The droppings shown in d, e, and f of figure 190 are of the Mormon cricket, caterpillar, and grasshopper, respectively, with enlargements to show the minute structure. In f the grasshoppers had fed on vegetation that produced as droppings packets of stiff bristlelike fibers neatly stacked together. In 1934 grasshoppers were extremely numerous on the dry lands along the Missouri in Montana, where these specimens were collected. The ground was strewn with the droppings that looked like seed heads of grass scattered about.

The caterpillar droppings in e are peculiar in being longitudinally segmented into hexagonal form, as shown in the cross section.

One day outside our kitchen window, under some tall plants, I found a scattering of these crinkly pellets, or frass, as caterpillar droppings are called. Searching among the leaves directly above I found the large caterpillar industriously chewing at the border of a leaf. Such frass can be a guide for the entomologist, not only to locate the caterpillar but to identify the insect, once a more complete knowledge of such material can be classified. A good method toward that end is to capture an insect or caterpillar, then collect and record by drawing and description the droppings or frass deposited in the cage soon after capture. By this means I obtained positive identification of some of the samples shown in the accompanying figures.

In addition, there are the little drops of pitch on conifer bark, the little tufts of sawdust here and there—the many indications that mark the hidden work of bark beetles and other insects.

The world of insects has much room for exploration.

When my brother and I were in school in western Minnesota we used to find small holes in the muddy bottom of the Red River. Always in pairs, they were in the shallows near shore and were big enough for us to put our fingers into. Quite often when we poked a finger into one hole a crayfish would shoot out of the other end of the tunnel, tail first. After we had learned about this, and wanted to catch the crayfish to have a look at it just for fun, we put a finger in each entrance, got the crayfish between them, and lifted it out through the soft mud.

Later, along slough banks in North Dakota and in other places, I found the little mud chimneys by the shore, not far from the water, and learned that these mounds, with a vertical tunnel down through the center, were also constructed by crayfish.

Sand crabs have burrows on the ocean beach and, as reported on page 114, on the east coast of Florida such burrows may be confused with the burrows of the oldfield mouse, *Peromyscus polionotus,* that lives there.

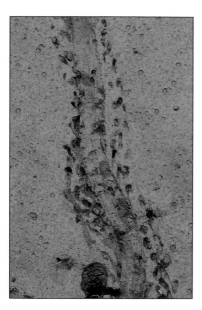

Through the shallow water along the banks of the Red River we also saw the freshwater clams and the trails or grooves they left in the mud as they walked along on the fleshy "foot" they put out through the partly opened shell. On certain sandy ocean beaches we are accustomed to locate marine clams at low tide by the slight mounds on the sand. Clam and crayfish sign are shown here in figure 188.

A crayfish trail in southern New Hampshire. Trail width was 1 in.

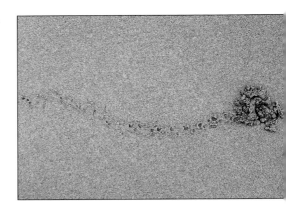

The trail and location of a mole crab on an Oregon beach. The width of the trail was ½ in.

Everything that crawls makes a track sooner or later, somewhere. You may have noticed an unusual number of earthworms about after a rain, sometimes crawling over the sidewalk. It has been said that they fall with the rain. The fact is that after a heavy rain earthworms tend to come out to the surface of the ground, and in muddy spots you may see the network of their trails, such as those shown in figure 193. If you look closely you will notice that in the softest mud their trails are the widest, over ⅛ of an inch. When they come to firmer ground their trail is narrower (they don't sink as

Figure 188. Crayfish and clam sign
a. Hollow dirt column thrown up by crayfish on shore.
b. Crayfish burrow with two entrances, in mud under water.
c. Clam and its trail in mud.

deep there), and where they strike drier spots you will find gaps in the trails, where the soft bodies made no impressions.

You will also find holes in the mud, with the diameter of the wider portions of the trail, where the earthworms came out for air. At some of these holes, or covering them, will be a tiny pile of very small soft mud pellets left there by the worm (see figure 193).

Figure 189. Insect sign

a. Runway of mole cricket, *Gryllotalpa*.
b. Mound of earth thrown up by a burrowing beetle of the genus *Geotrupes*. Mound sometimes has diam. of 6 in. Another mound-maker, *Bolboceras*, shown at right.
c. Centipede tracks and, at right, droppings of millipede, *Spirobolus* (Florida).
d. Droppings of katydid, *Pterophylla*.

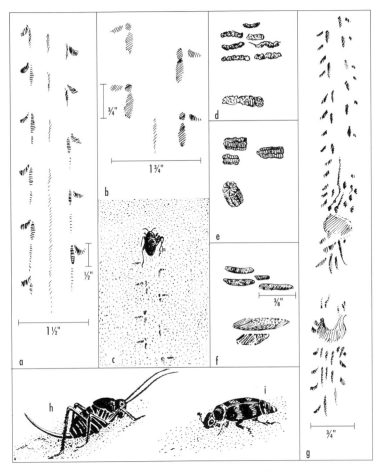

Figure 190. Mormon cricket, beetle, and grasshopper

a and b. Tracks of Mormon cricket in dust (Wyoming).

c. Beetle leaving tracks in desert sand.

d. Droppings of Mormon cricket, about natural size; enlarged figure below shows structure.

e. Caterpillar frass, probably one of the swallowtails; enlarged figure below shows hexagonal cross section.

f. Grasshopper droppings, about natural size, with enlarged samples below.

g. Grasshopper trail in sand; a shuffle with one short hop.

h. Mormon cricket.   i. Carrion beetle.

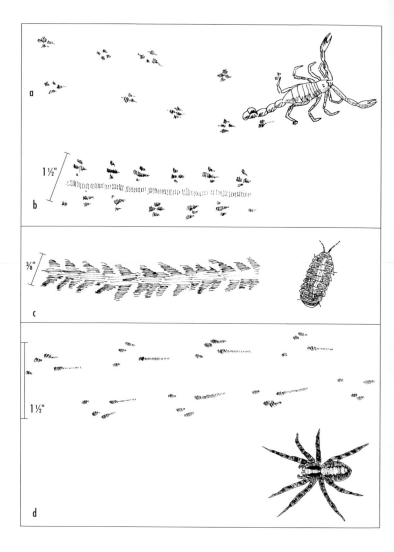

Figure 191. More invertebrate tracks
a. Scorpion with its tail up in defensive position.
b. Scorpion traveling with tail down.
c. Sow bug, or pill bug, trail.
d. Wolf spider.

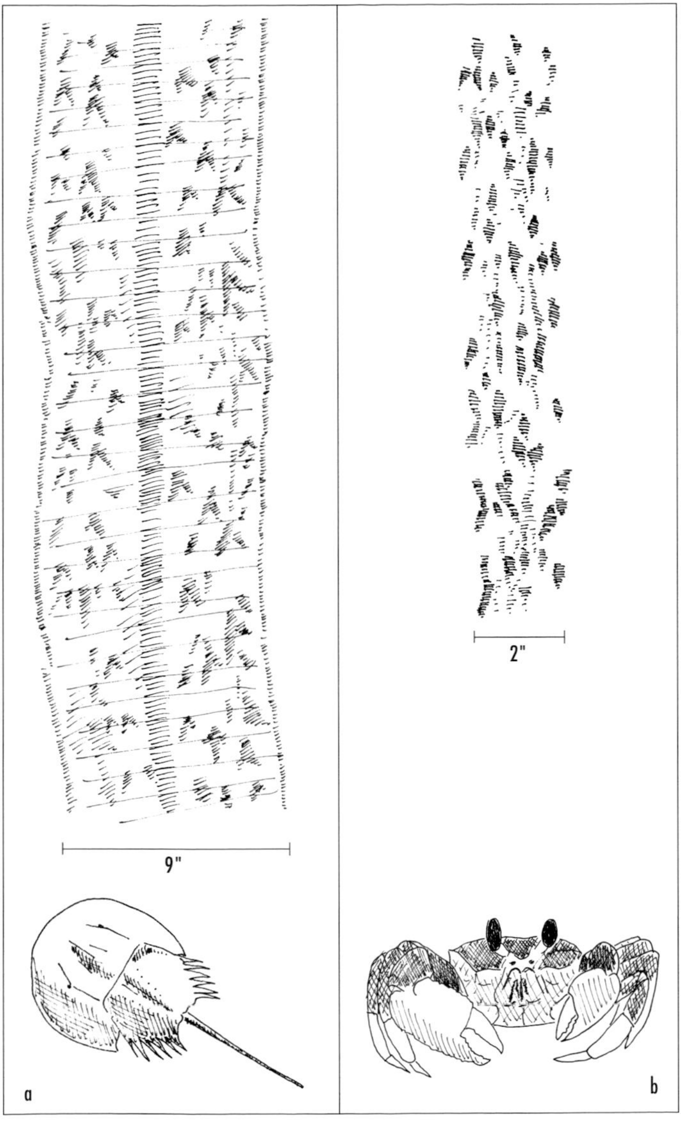

Figure 192. Seaside invertebrates

a. Horseshoe crab. b. Ghost crab.

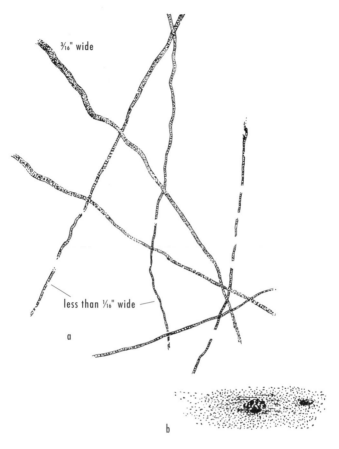

¾₁₆" wide

less than ⅟₁₆" wide

a

b

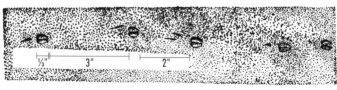

⅓" 3" 2"

c

Figure 193

a. Earthworm trails in wet mud.
b. Earthworm burrows, one plugged by tiny mud pellets.
c. Mysterious tracks.

TWIGS AND LIMBS

ANIMAL LIFE depends ultimately on vegetation for existence, and it is to be expected that shrubs and trees should furnish a share of this nourishment. One spring in Wyoming, when the snow had melted away, I found an extensive patch of buffalo berry bushes in a river bottom, conspicuously encircled by a light-colored band where meadow voles had eaten the bark. Often you will find the bark of wild rosebushes eaten, under the snow, and the stems cut into lengths (figure 194, a). The tooth marks are tiny, less than ¹⁄₁₆ of an inch.

One fall in Wyoming an early snowfall came while the aspens and cottonwoods still had their leaves. As a result, they held so much snow that the woods were filled with broken trees and limbs. Next spring when the snow melted away we found that the

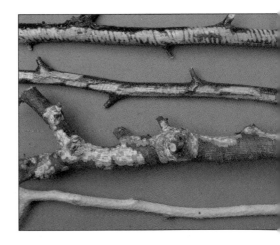

From top: a yellow birch chewed by a porcupine in New Hampshire; a beaked hazelnut chewed by a snowshoe hare in Maine; an alder chewed by a dusky-footed woodrat in California; a bigtooth aspen twig completely debarked by a meadow vole in Vermont.

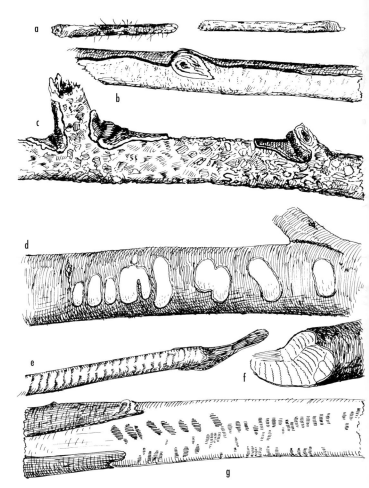

Figure 194. Gnawed twigs and limbs

a. Wild rose twigs gnawed by meadow voles in winter. Note that the slanted cut at the end is not done by a single bite, but by successive tooth cuts. See also figure 57, b.

b. Cottonwood limb gnawed by meadow voles under the snow.

c. Cottonwood limb gnawed by pocket gophers under the snow.

d. Aspen limb scraped by a moose's lower incisors. See also figure 152.

e. Bark eaten off willow twig by beaver.

f. End of willow stem severed by beaver.

g. Beaver gnawings on a cottonwood limb. See also figure 46, c.

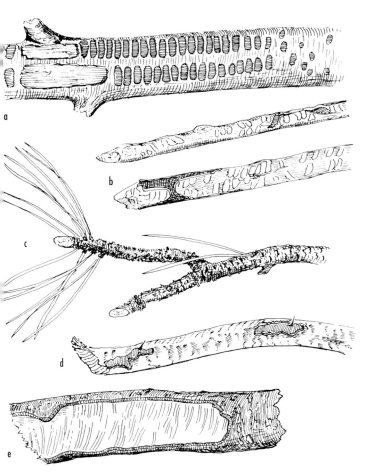

Figure 195. Gnawed twigs and limbs

a. Pits of sapsucker in a willow stem. Pits are up to ⅛ in. wide.

b. Cottontail gnawings. Tooth marks vary from a little less than ¹⁄₁₆ to a little less than ⅛ in. (Wisconsin).

c. Tips of lodgepole pine nipped by snowshoe hare (Wyoming). See also illustration on p. 366.

d. Porcupine gnawing of bark. Smaller tooth marks up to ¹⁄₁₆ in., larger ones ⅛ in. wide (Idaho).

e. Very old gnawings of porcupine on spruce limbs (Wyoming). Tooth marks about ¹⁄₁₆ in. wide.

dwellers under the deep snow had fared very well indeed. Cotton-wood and aspen bark is palatable to many animals, and the fallen limbs everywhere indicated evidence of winter feasting. Figure 194, c, shows cottonwood-limb markings left by pocket gophers, which had tunneled through the snow and left some of their earth cores over and beside the limbs. Note the smooth surface of the vole work in b and the coarse appearance of c, where the pocket gophers had eaten the bark and also gouged into the wood. The gopher tooth marks were about 1/16 of an inch, some smaller, others considerably larger.

Moose, elk, and deer scrape bark with their lower incisors, and this sign can often be found on aspen, fir, red maple, hemlock, witch hazel, and occasionally alder and willow. Usually the gnawings are vertical or diagonal stripes on the trunk (see figures 137 and 152). Occasionally on a slanted or horizontal large limb you will find a few moose bites, as in figure 194, d. These marks measure from 1/4 to 1/2 inch in width.

Of course the master lumberman is the beaver, who sometimes fells trees as large as 3 feet in diameter. Figure 194, e, shows a

Figure 196.
Tree borings of the pileated wood-pecker. These are made by the bird when seeking out ant colonies in the trunks of trees or stumps. The openings may be in many sizes, some upward of a foot in length, others small, very often in groups as shown here. Generally they are elongated, and squared off at the ends, but there are variations in size and shape, all depending on the way the bird adapts its work to the need for reaching the insects within.

It is interesting to notice that very often the holes drilled by the sapsucker also tend to be square or oblong in shape; see figure 195, a.

Figure 197

a. An old blaze drawn by Paul Brooks (Lincoln, Massachusetts). This mark is often hard to tell from the scar left by sawing off a branch.

b. Aspen trunk with an old, healed lightning strike; a blaze mark has similar rough edges but is much shorter in length.

twig stripped of bark; the tooth marks here are about ⅛ inch wide. In f the beaver has severed a willow stem, leaving tooth marks about ⅛ to ¼ inch wide, while g shows a cottonwood limb stripped of bark by a beaver, with a few gouges in the wood from ¹⁄₁₆ and ⅛ to ¼ inch in width. Where beavers have been working you will also find beaver gnawings on the tree itself, bark being removed as high as 6 feet above the ground. The explanation of this is that the beaver was out working on the surface of deep snow.

Sapsuckers leave neat rows of pits in the bark (see figures 195, a, and 196), while other woodpeckers drill holes of varying size, and there are many other scars left on trees by insects or accidents. Lightning marks are among the most common of these. Lightning leaves long vertical streaks on trees, often somewhat spiral and generally neatly healed over on aspens. At other times you will find trees broken and splintered by lightning.

A mark that all woodsmen must learn to recognize is the "blaze" (see figure 197, a), made by slashing a small patch in the tree's bark with an ax. This is a sign left by a human to indicate a trail through the forest. Knowledge of it can often save you from an unexpected night outdoors.

Figure 198.
Spruce grouse and blue grouse both feed on spruce and fir twigs, and leave partially denuded twigs such as shown here.

Snowshoe hares sometimes leave twigs in this same condition.

BONE AND HORN CHEWING

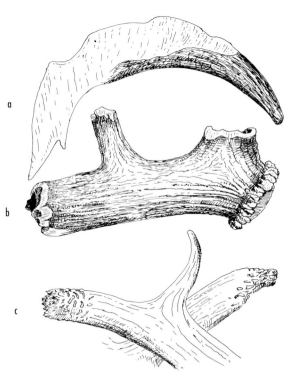

Figure 199. Horn-chewing sign
a. All that is left of a mountain sheep horn eaten by rodents.
b. All that is left of a large elk antler, almost completely consumed by rodents.
c. Tips of two elk antlers chewed by elk.

ONE WINTER DAY when I drove by dog team to a reindeer herd in the Kuskokwim country of Alaska, I saw intensified to a remarkable degree a habit that is fairly common among the deer. We know they like to nibble at antlers. The reindeer I saw were not only chewing the shed antlers, but were eating them off the owners' heads! Many of those who still carried their antlers had them chewed down to smaller size by their neighbors. The females, especially, were feeding on the spike antlers of their fawns. Antler chewing has been noted in several places among the deer, and the habit is very prevalent among the Wyoming elk herds. Such chewing is done by the molariform teeth, or grinders, and the rough marks made by those teeth are distinctive. Note the antler tips in figure 199, c, with their appearance of having been gouged and "worn" off in a clumsy manner.

Note by contrast the neat carving effect in b, the work of rodents equipped with gnawing teeth. These rodents no doubt include mice, voles, ground and tree squirrels, and the porcupine. I have not been able to identify properly the specific rodents who do this work, but I do know that they are extremely efficient. Figure 199, b, shows all that remained of a large elk antler, and a, the nearly finished horn of a mountain sheep. I also found a large leg bone of a horse well trimmed down.

Wolves and coyotes also chew on various antlers, and will eat moose antlers almost completely. They especially savor the softer portions of moose antlers, found in the center between the tines.

All bones are chewed by rodents. Here a deer bone and a raccoon skull show severe signs of feeding rodents.

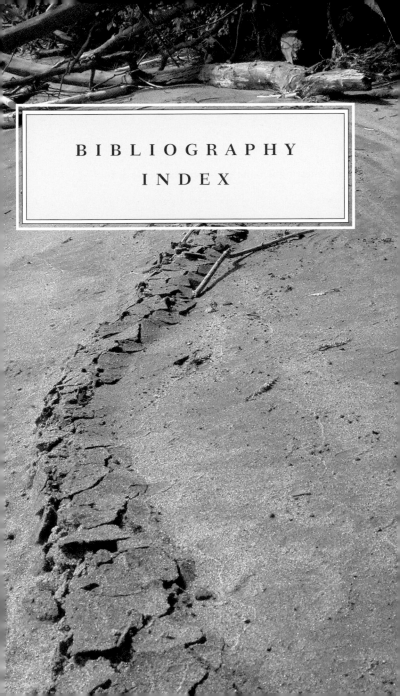

BIBLIOGRAPHY

INDEX

BIBLIOGRAPHY

ken, Russell. "Tracks: a unique method of collecting them." *Nature* 15 (3):170 (1930). Describes plaster cast method.

corn, J. R. "On the decoying of coyotes." *Journal of Mammalogy* 27 (2): 122–26 (1946).

ien, A. A. "A Christmas walk with birds and beasts." *American Forestry* 25 (312): 1526–30 (1919).

ien, Durward L. "Notes on the killing technique of the New York weasel." *Jour. Mamm.* 19 (2):225–29 (1938).

nthony, Harold E. "The bat." *Natural History* 25 (6):560–70 (1925).

——. *Field book of North American mammals.* Putnam, 1928.

anda, M. *Huellas y otros rastros de los mamíferos grandes y medianos de México.* Xalapa, Mexico: Instituto de Ecología, 2000.

lton, A.V. "An ecological study of the mole." *Jour. Mamm.* 17 (4):349–71 (1936).

udubon, John James, and John Bachman. *The quadrupeds of North America.* V. G. Audubon, 1851–54.

iley, Vernon. "Biological survey of Texas." *U.S. Dept. of Agric., Bur. of Biol. Survey No. Am. Fauna* 25 (1905).

——. "Capturing small mammals for study." *Jour. Mamm.* 2 (2):63–68 (1921).

——. "Mammals of New Mexico." *U.S. Dept. of Agric., Bur. of Biol. Survey No. Am. Fauna* 53 (1931).

——. "Trapping animals alive." *Jour. Mamm.* 13 (4):337–42 (1932).

ker, R., et al. *Revised checklist of North American mammals north of Mexico,* 2003. (No. 229) Mus. of Texas Tech Univ., 2003.

nson, Seth B. "Decoying coyotes and deer." *Jour. Mamm.* 29 (4):406–9 (1948).

nson, Seth B., and Adrey E. Borell. "Notes on the life history of the red tree mouse, *Phenacomys longicaudus.*" *Jour. Mamm.* 12 (3):226–33 (1931).

Bergtold, W. H. "Unusual nesting of a raccoon." *Jour. Mamm.* 6 (4):280–81 (1925).

Blair, W. Frank. "Notes on home ranges and populations of the short-tailed shrew." *Ecology* 21 (2):284–88 (1940).

———. "Some data on the home ranges and general life history of the short-tailed shrew, red-backed vole and woodland jumping mouse in northern Michigan." *American Midland Naturalist* 25 (3):681–85 (1941).

Brown, Tom. *Tom Brown's field guide to nature observation and tracking.* Berkeley, 1983.

Brunner, Josef. *Tracks and tracking.* Outing Pub. Co., 1909.

Burt, William H. *The mammals of Michigan.* Univ. of Michigan Press, 1946.

———. "A simple live trap for mammals." *Jour. Mamm.* 8 (4):302–4 (1927).

Burt, William H., and Richard P. Grossenheider. *A field guide to the mammals.* 3rd ed. Houghton Mifflin, 1975.

Cahalane, Victor H. *Mammals of North America.* Macmillan, 1947.

Childs, J. *Tracking the felids of the borderlands.* Self-published, 1998. (To order, call 520-883-4029.)

Chubb, S. Harmsted. "How animals run." *Natural History* 29 (5):543–51 (1929).

Clark, Harold W. "Records of night: the ramblers are revealed." *Nature* 14 (4):222 (1929).

Crabb, Wilfred D. "The ecology and management of the prairie skunk in Iowa." *Ecological Monographs* 18:201–32 (1948).

———. "Food habits of the prairie spotted skunk in southeastern Iowa." *Jour. Mamm.* 22 (4):349–64 (1941).

Dalquest, Walter W. "Mammals of Washington." *Mus. of Natural History, Univ. of Kansas Publ.* 2:1–444 (1948).

Davis, William B., and Leonard Joeris. "Notes on the life history of the little short-tailed shrew." *Jour. Mamm.* 26 (2):136–38 (1945).

Davis, William B., and Walter P. Taylor. "The bighorn sheep of Texas." *Jour. Mamm.* 20 (4):440–55 (1939).

de la Croix, P. Magne. "The evolution of locomotion in mammals." *Jour. Mamm.* 17 (1):51–54 (1936).

Dixon, Joseph S. "Notes on the life history of the gray shrew." *Jour. Mamm.* 5 (1):1–6 (1924).

Eadie, W. Robert. "A contribution to the biology of *Parascalops breweri.*" *Jour. Mamm.* 29 (2):150–73 (1939).

Elbroch, Mark. *Animal tracks: A "quick-guide" to animal track identification.* Redwood City, Cal.: Local Birds Inc., 2004

———. *Mammal tracks and signs: A guide to North American species.* Stackpole Books, 2003.

broch, Mark, with Eleanor Marks. *Bird tracks and signs: A guide to North American species.* Stackpole Books, 2001.

rington, Paul L. "An analysis of mink predation upon muskrats in north-central United States." *Iowa State Coll. of Agric. Research Bull.* No. 320 (1943).

y, Walter. "The wolverine." *Cal. Fish and Game* 9 (4):129–34 (1923).

les, LeRoy W. "Food habits of the raccoon in eastern Iowa." *Jour. Wildlife Management* 4 (4):375–82 (1940).

lass, Bryan P. "The black-footed ferret in Oklahoma." *Jour. Mamm.* 31 (4):460 (1950).

lover, Fred A. "Killing techniques of the New York weasel." *Penn. Agric. Exp. Sta.*, Paper No. 1147 (1942).

——. "A study of the winter activities of the New York weasel." *Penn. Game News* 14 (6):8–9 (1943).

rinnell, George Bird. "Mountain sheep." *Jour. Mamm.* 9 (1):1–9 (1928).

rinnell, Joseph. "Bats as desirable citizens." *Cal. Fish and Game Commission, Teachers' Bull.* No. 6 (1916).

——. "Disgorgement among songbirds." *Auk* 14 (4):412 (1897).

rinnell, Joseph, Joseph S. Dixon, and Jean M. Linsdale. *Fur-bearing mammals of California.* Univ. of California Press, 1937.

alfpenny, James. *A field guide to mammal tracking in western America.* Johnson Books, 1986.

all, E. Raymond. "Breeding habits of the short-tailed shrew, *Blarina brevicauda.*" *Jour. Mamm.* 10 (2):125–34 (1929).

——. "Food of the Soricidae." *Jour. Mamm.* 11 (1):26–39 (1930).

——. "Habits of the star-nosed mole, *Condylura cristata.*" *Jour. Mamm.* 12 (4):345–55 (1931).

——. *Mammals of Nevada.* Univ. of California Press, 1946.

——. "Notes on the life history of the sage-brush meadow mouse (*Lagurus*)." *Jour. Mamm.* 9 (3):201–4 (1928).

——. "A revised classification of the American ermines with description of a new subspecies from the western Great Lakes region." *Jour. Mamm.* 26 (2):175–82 (1945).

——. "The weasels of New York: their natural history and economic status." *Am. Midland Naturalist* 14 (4):289–344 (1933).

amilton, William J., Jr. "Activity of Brewer's mole (*Parascalops breweri*)." *Jour. Mamm.* 20 (3):307–10 (1939).

——. *American mammals.* McGraw-Hill, 1939.

——. "The biology of the little short-tailed shrew *Cryptotis parva.*" *Jour. Mamm.* 25 (1):1–7 (1944).

——. "The biology of the smoky shrew." *Zoologica* 25 (4):473–92 (1940).

————. "Exploring the world of 'whistle pig.'" *Audubon* 52 (2):96–101 (1950).

————. "Life history notes on the northern pine mouse." *Jour. Mamm.* 19 (2):163–70 (1938).

Handley, Charles O., Jr., and Clyde P. Patton. *Wild mammals of Virginia.* Va. Commission of Game and Inland Fisheries, 1947.

Harper, Francis. "The Florida water-rat *(Neofiber alleni)* in Okefenokee Swamp, Georgia." *Jour. Mamm.* 1 (2):65–66 (1920).

Hartman, Carl G. *Possums.* Univ. of Texas Press, 1952.

Herlocker, Emmett. "Mountain beaver—biological curiosity." *Audubon* 52 (6):387–90 (1950).

Hickie, Paul. "New developments in a small mammal trap." *Jour. Mamm.* 16 (1):71–73 (1935).

Hisaw, Frederick L. "Observations on the burrowing habits of moles *(Scalopus aquaticus machrinoides)*." *Jour. Mamm.* 4 (2):79–88 (1923).

Hooper, Emmet T. "The water shrew of the southern Allegheny Mountains." *Mus. of Zoology, Univ. of Michigan,* Occasional Papers No. 463 (1942).

Howell, Arthur H. "Description of a new race of the Florida water-rat *(Neofiber alleni)*." *Jour. Mamm.* 1 (2):79–80 (1920).

————. "A revision of the American Arctic hares." *Jour. Mamm.* 17 (4):315–37 (1936).

Ingles, Lloyd Glenn. *Mammals of California.* Stanford Univ. Press, 1946.

Ivey, R. DeWitt. "Life history notes on three mice from the Florida east coast." *Jour. Mamm.* 30 (2):157–62 (1949).

Jackson, Hartley H. T. "A review of the American moles." *U.S. Dept. of Agric., Bur. of Biol. Survey No. Am. Fauna* 38 (1915).

————. "A taxonomic review of the American long-tailed shrews." *U.S. Dept. of Agric., Bur. of Biol. Survey No. Am. Fauna* 51 (1928).

Jaeger, Ellsworth. *Tracks and trailcraft.* Macmillan, 1948.

Jewett, Stanley G. "Notes on two species of *Phenacomys* from Oregon." *Jour. Mamm.* 1 (4):165–68 (1920).

Johnson, C. Stuart. "Tracks from the Pleistocene of west Texas." *Am. Midland Naturalist* 18 (1): 147–52 (1937).

Kays, Roland W., and Don E. Wilson. *Mammals of North America.* Princeton Univ. Press, 2002.

Klugh, A. Brooker. "Notes on the habits of *Blarina brevicauda.*" *Jour. Mamm.* 2 (1):35 (1921).

Lang, Herbert. "Position of limbs in the sliding otter." *Jour. Mamm.* 5 (3):216–17 (1924).

Lay, Daniel W. "Ecology of the opossum in eastern Texas." *Jour. Mamm.* 23 (2):147–59 (1942).

Liebenberg, Louis. *The art of tracking: the origin of science.* Cape Town, South Africa: David Philip, 1990.

ers, Emil E. "Notes on the river otter *(Lutra canadensis)*." *Jour. Mamm.* 32 (1):1–9 (1951).

ord, R. D., A. M. Vilches, J. I. Maiztequi, and C. A. Soldini. "The trackingboard: a relative census technique for studying rodents." *Jour. Mamm.* 51 (4):828 (1970).

ucas, Frederick A. "Fossil footprints." *Evolution* 10:5 (1928).

arshall, William H. "Mink displays sliding habits." *Jour. Mamm.* 16 (3):228–29 (1935).

———. "A study of the winter activities of the mink." *Jour. Mamm.* 17 (4):382–92 (1936).

ason, George F. *Animal tracks.* Morrow, 1943.

cCabe, Robert A. "Notes on live-trapping mink." *Jour. Mamm.* 30 (4):416–23 (1949).

cLean, Donald D. "The prong-horned antelope in California." *Cal. Fish and Game* 30 (4):221–41 (1944).

ills, Enos A. *The grizzly.* Houghton Mifflin, 1919.

oore, A. W. "Improvements in live trapping." *Jour. Mamm.* 17 (4):372–74 (1936).

———. "Notes on the sage mouse in eastern Oregon." *Jour. Mamm.* 24 (2):188–91 (1943).

———. "Notes on the Townsend mole." *Jour. Mamm.* 20 (4):499–501 (1939).

urie, Adolph. "Cattle on grizzly bear range." *Jour. Wildlife Management* 12 (1):57–72 (1948).

———. "Ecology of the coyote in the Yellowstone." *U.S. Natl. Park Service, Fauna Ser.* No. 4 (1940).

———. "Following fox trails." *Mus. of Zoology, Univ. of Michigan, Misc. Publications* No. 32 (1936).

———. "Some food habits of the black bear." *Jour. Mamm.* 18 (2):238–40 (1937).

———. "The Wolves of Mt. McKinley." *U.S. Natl. Park Service, Fauna Ser.* No. 4 (1940).

urie, Olaus J. *The elk of North America.* Stackpole Co. and Wildlife Mgt. Inst, 1951.

———. "The mink—brown mischief." *Home Geographic Monthly* 1 (7):43–48 (1932).

———. "Notes on the sea otter." *Jour. Mamm.* 21 (2):119–31 (1940).

uybridge, Eadweard. *Animals in motion.* Dover Publications, 1957.

elson, Edward W. "Bats in relation to the production of guano and the destruction of insects." *U.S. Dept. of Agric., Dept. Bull.* No. 1395 (1926).

———. "Report upon natural history collections made in Alaska." U.S. Army Signal Service, 1887.

elson, Edward W., and Louis Agassiz Fuertes. *Wild animals of North America.* Natl. Geographic Society, 1930.

Newman, H. H. "The natural history of the nine-banded armadillo of Texas." *Am. Naturalist* 67 (561):513–39 (1913).

Palmer, E. Laurence. "Larger mammals." *Cornell Rural School Leaflet* 19 (2):1–44 (1925).

———. "A talk on winter animals." *Cornell Rural School Leaflet* 13 (3):83–130 (1920).

Pearce, John. "Identifying injury by wildlife to trees and shrubs in northeastern forests." *U.S. Fish and Wildlife Ser., Research Report* No. 13 (1947).

Pettigrew, J. Bell. *Animal locomotion.* Appleton, 1874.

Polderboer, Emmett B. "Habits of the least weasel *(Mustela rixosa)* in northeastern Iowa." *Jour. Mamm.* 23 (2):145–47 (1942).

Polderboer, Emmett B., Lee W. Kuhn, and George O. Hendrickson. "Winter and spring habits of weasels in central Iowa." *Jour. Wildlife Management* 5 (1):115–19 (1941).

Proctor, Thomas. "Disgorgement among songbirds." *Auk* 14 (4):412 (1897).

Quick, Edgar R., and A. W. Butler. "The habits of some Arvicolinae." *Am. Naturalist* 19 (2):113–18 (1885).

Quick, H. F. "Habits and economics of the New York weasel in Michigan." *Jour. Wildlife Management* 8 (1):71–78 (1944).

Reed, Charles A., and Thane Riney. "Swimming, feeding, and locomotion of a captive mole." *Am. Midland Naturalist* 30 (3):790–91 (1943).

Reid, F. *A field guide to the mammals of Central America and southeast Mexico.* Oxford Univ. Press, 1997.

Reynolds, Harold C. "Some aspects of the life history and ecology of the opossum in central Missouri." *Jour. Mamm.* 26 (4):361–79 (1905).

Rezendes, Paul. *Tracking and the art of seeing: How to read animal tracks and sign.* 2nd ed. HarperCollins Publishers, 1999.

Rollings, Clair T. "Habits, foods, and parasites of the bobcat in Minnesota." *Jour. Wildlife Management* 9 (2):131–45 (1945).

Rossell, Leonard. *Tracks and trails.* Boy Scouts of America, distr. by Macmillan, 1928.

Rutherford, Ralph L., and Loris S. Russell. "Mammal tracks from the Paskapoo beds in Alberta." *Am. Jour. Science* 5th ser., 15 (87):262–64 (1928).

Scheffer, Theophilus H. "The common mole of eastern United States." *U.S. Dept. of Agric., Farmers' Bull.* No. 583 (1917).

———. "Hints on live trapping." *Jour. Mamm.* 15 (3):197–202 (1934).

———. "Trapping moles and utilizing their skins." *U.S. Dept. of Agric., Farmers' Bull.* No. 832 (1917).

Schmidt, F. J. W. "Mammals of western Clark County, Wisconsin." *Jour. Mamm.* 12 (2):99–117 (1931).

cott, Thomas G. "The secrets of the trail." *Country Life in America* 8 (2):202–5 (1905).

———. "Some food coactions of the northern plains red fox." *Ecological Monographs* 13 (4):427–79 (1943).

———. "Stories on the tree-trunks." *Country Life in America* 6 (1):37–39, 90 (1904).

eton, Ernest Thompson. *Lives of game animals.* Doubleday, 1929.

———. "The mole-mouse, potato mouse, or pine-mouse." *Jour. Mamm.* 1 (4): 185 (1920).

———. "On the study of scatology." *Jour. Mamm.* 6 (1):47–49 (1925).

everinghouse, C. W., and John E. Tanck. "Speed and gait of an otter." *Jour. Mamm.* 29 (1):71 (1948).

haw, William T. "Alpine life of the heather vole (*Phenacomys olympicus*)." *Jour. Mamm.* 5 (1):12–15 (1924).

hull, A. Franklin. "Habits of the short-tailed shrew." *Am. Naturalist* 61 (488):495–522 (1907).

ilver, James. "The European hare (*Lepus europaeus* Pallas) in North America." *Jour. Agricultural Research* 28 (11):1133–37 (1924).

ilver, James, and A. W. Moore. "Mole control." *U.S. Fish and Wildlife Ser., Conservation Bull.* No. 16 (1941).

impson, Sutturland Eric. "The nest and young of the star-nosed mole." *Jour. Mamm.* 4 (3): 167–71 (1923).

kinner, M. P. *Bears in the Yellowstone.* McClurg, 1925.

mith, Clarence P. "Notes on the habits of the long-tailed harvest mouse." *Jour. Mamm.* 17 (3):274–78 (1936).

ooter, Clarence A. "Habits of coyotes in destroying nests and eggs of waterfowl." *Jour. Wildlife Management* 10 (1):33–38 (1946).

tegeman, LeRoy C. "The European wild boar in the Cherokee National Forest, Tennessee." *Jour. Mamm.* 19 (3):279–90 (1938).

tephainsky, H. "Mit der Kamera im Spurschnee." *Der Naturforscher* 6 (6): 211–17 (1929).

tickel, Lucille F., and William H. "A *Sigmodon* and *Baiomys* population in ungrazed and unburned Texas prairie." *Jour. Mamm.* 30 (2): 141–50 (1949).

vihla, Arthur. "Life history notes on *Sigmodon hispidus hispidus.*" *Jour. Mamm.* 10 (4):352–53 (1929).

———. "Life history of the Texas rice rats (*Oryzomys palustris texensis*)." *Jour. Mamm.* 12 (3):238–42 (1931).

———. "The mountain water shrew." *Murrelet* 15 (2):44–45 (1934).

vihla, Arthur, and Ruth D. Svihla. "Mink feeding on clams." *Murrelet* 12 (1):22 (1931).

vihla, Ruth D. "Habits of *Sylvilagus aquaticus littoralis.*" *Jour. Mamm.* 10 (4): 315–19 (1929).

aber, F. Wallace. "Contribution on the life history and ecology of the nine-banded armadillo." *Jour. Mamm.* 26 (3):211–26 (1945).

Tappe, Donald T. "Natural history of the Tulare kangaroo rat." *Jour. Mamm.* 22 (2):117–48 (1941).

Tevis, Lloyd P., Jr. "Summer activities of California raccoons." *Jour. Mamm.* 28 (4):323–32 (1947).

Walker, Alex. "Notes on the forest *Phenacomys*." *Jour. Mamm.* 11 (2):233–35 (1930).

Walker, Lewis Wayne. "Passers-by." *Outdoor Life* 70 (1):50–52 (1932).

Warren, Edward R. *The mammals of Colorado.* Putnam, 1910.

Whitney, Leon F. "The raccoon and its hunting." *Jour. Mamm.* 12 (1):29–38 (1931).

Wilson, Clifford. "Animal trackers." *Rod and Gun in Canada* 36 (8):20–21 (1935).

Wilson, Don E., and Sue Ruff. *The Smithsonian book of North American mammals.* Smithsonian Institute Press, 1999.

Wood, Rodney C. *Animal tracking for Boy Scouts: Hints on animal tracking.* Canadian General Council for Boy Scouts Assoc., 1924.

Wright, William H. *The grizzly bear.* Scribner, 1909.

Yeager, Lee E., and R. G. Reunels. "Fur yields and autumn foods of the raccoon in Illinois River bottomlands." *Jour. Wildlife Management* 7 (1):45–60 (1943).

Zielinski, W., and T. Kucera. *American marten, fisher, lynx and wolverine: survey methods for their detection.* Pacific Southwest Research Station, USFS, 1996.

INDEX

Page numbers in **boldface** indicate illustrations.

If I were to make a study
of the tracks of animals
and represent them by plates,
I should conclude
with the tracks of man.

—Henry David Thoreau

THE PETERSON SERIES®

PETERSON FIELD GUIDES®

BIRDS

ADVANCED BIRDING North America 97500-x
BIRDS OF BRITAIN AND EUROPE 0-618-16675-0
BIRDS OF TEXAS Texas and adjacent states 92138-4
BIRDS OF THE WEST INDIES 0-618-00210-3
EASTERN BIRDS Eastern and central North America 74046-0
EASTERN BIRDS' NESTS U.S. east of Mississippi River 93609-8
HAWKS North America 67067-5
HUMMINGBIRDS North America 0-618-02496-4
WESTERN BIRDS North America west of 100th meridian and north of Mexico 91173-7
WESTERN BIRDS' NESTS U.S. west of Mississippi River 0-618-16437-5
MEXICAN BIRDS Mexico, Guatemala, Belize, El Salvador 97514-x
WARBLERS North America 78321-6

FISH

PACIFIC COAST FISHES Gulf of Alaska to Baja California 0-618-00212-x
ATLANTIC COAST FISHES North American Atlantic coast 97515-8
FRESHWATER FISHES North America north of Mexico 91091-9

INSECTS

INSECTS North America north of Mexico 91170-2

BEETLES North America 91089-7
EASTERN BUTTERFLIES Eastern and central North America 90453-6
WESTERN BUTTERFLIES U.S. and Canada west of 100th meridian, part of northern Mexico 79151-0

MAMMALS

MAMMALS North America north of Mexico 91098-6
ANIMAL TRACKS North America 0-618-51743-x

PETERSON FIELD GUIDES®

ECOLOGY

PLANTS

EARTH AND SKY

PETERSON FIELD GUIDES® continued

REPTILES AND AMPHIBIANS

EASTERN REPTILES AND AMPHIBIANS Eastern and
central North America 90452-8

WESTERN REPTILES AND AMPHIBIANS Western North
America, including Baja California 93611-x

SEASHORE

SHELLS OF THE ATLANTIC Atlantic and Gulf coasts
and the West Indies 0-618-16439-1

PACIFIC COAST SHELLS North American Pacific coast, including
Hawaii and the Gulf of California 18322-7

ATLANTIC SEASHORE Bay of Fundy to Cape Hatteras 0-618-00209-x

CORAL REEFS Caribbean and Florida 0-618-00211-1

SOUTHEAST AND CARIBBEAN SEASHORES Cape Hatteras to the Gulf
Coast, Florida, and the Caribbean 97516-6

PETERSON FIELD GUIDE AUDIOS

EASTERN BIRDING BY EAR
cassettes 0-618-22591-9
CD 0-618-22590-0

MORE EASTERN BIRDING BY EAR
cassettes 97529-8
CD 97530-1

WESTERN BIRDING BY EAR
cassettes 97526-3
CD 97525-5

BACKYARD BIRDSONG
cassettes 0-618-22593-5
CD 0-618-22592-7

EASTERN BIRD SONGS, Revised
CD 0-618-22594-3

WESTERN BIRD SONGS, Revised
CD 97519-0

PETERSON FIELD GUIDE
COLOR-IN BOOKS

BIRDS 0-618-30722-2
BUTTERFLIES 0-618-30723-0
REPTILES AND AMPHIBIANS 0-618-30737-0
WILDFLOWERS 0-618-30735-4
MAMMALS 0-618-30736-2
DINOSAURS 0-618-54224-8

PETERSON FLASHGUIDES™

ATLANTIC COASTAL BIRDS 79286-x
PACIFIC COASTAL BIRDS 79287-8
EASTERN TRAILSIDE BIRDS 79288-6
WESTERN TRAILSIDE BIRDS 79289-4
HAWKS 79291-6
BACKYARD BIRDS 79290-8
TREES 82998-4
ANIMAL TRACKS 82997-6
BUTTERFLIES 82996-8
ROADSIDE WILDFLOWERS 82995-x
BIRDS OF THE MIDWEST 86733-9
WATERFOWL 86734-7

PETERSON FIELD GUIDES can be purchased at your local
bookstore or by calling our toll-free number, (800) 225-3362.
Visit **www.petersononline.com** for more information.
When referring to title by corresponding ISBN number,
preface with 0-395, unless title is listed with 0-618.